Power System Optimization Modeling in GAMS

Alireza Soroudi

Power System Optimization Modeling in GAMS

 Springer

Alireza Soroudi
School of Electrical Engineering
University College of Dublin
Belfield, Dublin, Ireland

ISBN 978-3-319-62349-8   ISBN 978-3-319-62350-4   (eBook)
DOI 10.1007/978-3-319-62350-4

Library of Congress Control Number: 2017944921

Printed on acid-free paper

This Springer imprint is published by Springer Nature
The registered company is Springer International Publishing AG
The registered company address is: Gewerbestrasse 11, 6330 Cham, Switzerland

Preface

The complex structure of electric power systems as the largest human-built machines justifies the need for an efficient and robust computation tool to conduct in-depth analysis. This book is intended to serve as an introduction to the application of GAMS software in solving a broad range of power system optimization problems. Due to the innate capability of GAMS, researchers can now pursue questions that they were unable to previously. It has been 8 years since I was introduced to GAMS. During these years, I have explored new areas in power system studies. The main purpose of this book is to share these experiences with other researchers/students or industry employees who might need a powerful computation tool for their studies. This book covers a broad range of topics in power system studies as follows:

- Chapter 1: Gives an overview and tips on how to start coding with GAMS for beginners.
- Chapter 2: Simple but practical examples are solved.
- Chapter 3: Economic dispatch concept is covered and solved.
- Chapter 4: Deals with dynamic economic dispatch problem where the problem is multi-period.
- Chapter 5: Covers the unit commitment problem.
- Chapter 6: Multi-period optimal DC/AC power flow is explained and solved.
- Chapter 7: The application of energy storage system in power system planning and operation is analyzed.
- Chapter 8: The power system observability is enhanced using PMU technology.
- Chapter 9: Some different transmission network operation and planning problems are presented and solved.
- Chapter 10: Energy system integration is discussed and solved.

The intended audiences are but not limited to the following:

- Power engineers: The proposed framework of this book enables power engineers to obtain a better insight into power system studies.
- Educators: This book can be used as a textbook in different power system operation and planning modules.

- System operators: Power system operators usually use commercial packages for solving their problems. This book can bring flexibility into their studies. Sometimes, the commercial software does not include all required analysis, or they are used as a black box.
- Researchers: Students and researchers with diverse mathematical and power system backgrounds can use this book to get familiar with different power system optimization issues.

This book is provided by the author "as is," and any express or implied warranties, including, but not limited to, the implied warranties of merchantability and fitness for a particular purpose, are disclaimed. In no event shall the copyright holder or author or publisher be liable for any direct, indirect, incidental, special, exemplary, or consequential damages (including, but not limited to, procurement of substitute goods or services; loss of use, data, or profits; or business interruption) however caused and on any theory of liability, whether in contract, strict liability, or tort (including negligence or otherwise) arising in any way out of the use of this software or codes, even if advised of the possibility of such damage.

As the first book on "power system optimization modeling in GAMS," there is always space for improvement. Any comments and suggestions from the readers are welcome. Please kindly share them with the author.

Dublin, Ireland   Alireza Soroudi
May 2017

Acknowledgments

This book would have not been finished without the contributions of several globally recognized experts in the power system field. A special thanks to Dr. Turaj Amraee who introduced GAMS to me for the first time.

I would like to thank those who gave me insight about the power system studies, especially Dr. Raphael Caire at Grenoble INP, Dr. Andrew Keane at University College Dublin, and Dr. Jonathon O'Sullivan at EirGrid.

This book is dedicated to my lovely wife, Soudeh, who supported me in putting together all the details. Finally, I appreciate all that my parents, Simin and Shahryar, and my sister, Mona, have done for me.

Contents

List of Figures

List of Tables

Chapter 1
Introduction to Programming in GAMS

The General Algebraic Modeling System (GAMS) is a modeling tool for mathematical programming and optimization purpose. This chapter provides the instruction on different programming elements in GAMS. It can be used in solving different types of optimization problems. Some basic optimization models used in power system literature are described in this chapter.

1.1 Optimization Problems in Power System

The power system optimization problems are broadly categorized as operation and planning problems. The operation problems are usually related to how to exploit the existing devices/power plants. For example, optimal power flow is an operation problem. The planning problems usually refer to those problems which investigate whether to invest or not in some assets. For example, the decisions regarding the transmission network expansion belong to planning category. Some of these problems are shown in Fig. 1.1. Some power system planning problems are listed as follows:

- Generation expansion planning (GEP) [1, 2]: In GEP, the decision maker is trying to find out the investment decision regarding the generation technology, size, and time of investment.
- Transmission expansion planning (TEP) [3, 4]. In TEP, the decision maker is trying to find out the investment decision regarding the planning option and time of investment. The planning options include but not limited to: building new lines, reconductoring the existing lines, building, or reinforcing the substations.
- Distribution network planning [5]: this problem is trying to make smart investments in new feeders/substations to meet the technical constraints.

© Springer International Publishing AG 2017
A. Soroudi, *Power System Optimization Modeling in GAMS*,
DOI 10.1007/978-3-319-62350-4_1

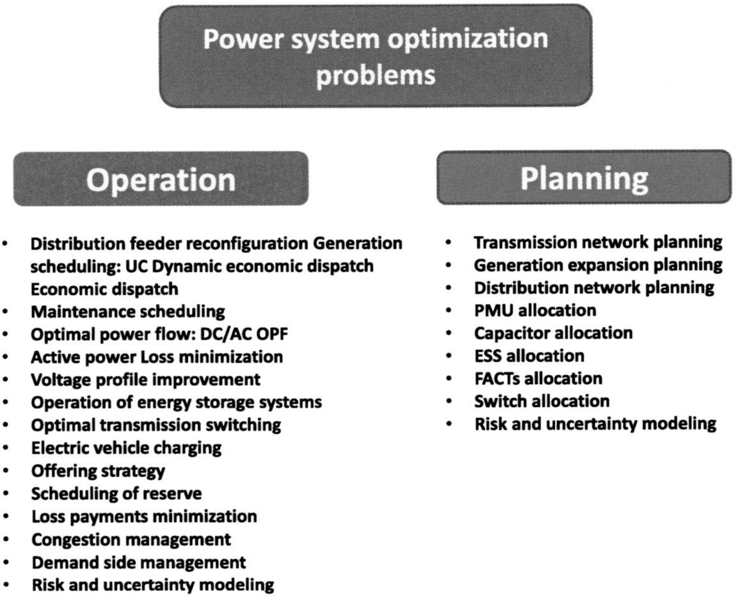

Power system optimization problems

Operation

- Distribution feeder reconfiguration Generation scheduling: UC Dynamic economic dispatch Economic dispatch
- Maintenance scheduling
- Optimal power flow: DC/AC OPF
- Active power Loss minimization
- Voltage profile improvement
- Operation of energy storage systems
- Optimal transmission switching
- Electric vehicle charging
- Offering strategy
- Scheduling of reserve
- Loss payments minimization
- Congestion management
- Demand side management
- Risk and uncertainty modeling

Planning

- Transmission network planning
- Generation expansion planning
- Distribution network planning
- PMU allocation
- Capacitor allocation
- ESS allocation
- FACTs allocation
- Switch allocation
- Risk and uncertainty modeling

Fig. 1.1 Some optimization problems in power system studies

- FACTS device allocation [6]: This problem is a subset of transmission planning problem. The FACTS devices can control the power flowing through a line and make the transmission network more flexible.
- Distributed generation (DG) allocation in distribution networks [7–9]: The optimal location (or size) of DG unit in a distribution network.
- Capacitor allocation in distribution networks [10]: The capacitor allocation is usually done at the distribution level. The purpose is usually improving the voltage profile in the network.
- PMU allocation [11, 12]: The system observability is improved using optimal placement of phasor measurement units in the network.
- Energy Storage System (ESS) allocation [13, 14]: The ESS can deliver value to customers. This highly depends on how they are operated and located in the system. This problem tries to maximize its benefits by finding the optimal connection point of ESS to the grid.
- Risk and uncertainty modeling in planning studies: the decision-making process highly relies on accuracy of input data. As a matter of fact, the input data for any decision-making problem (especially in practical problems) are subject to uncertainty. If these uncertainties are not treated and handled properly, then costly consequences might occur. Based on the data availability some of the following techniques are applicable:

    - Information gap decision theory (IGDT): ESS allocation [15]
    - Scenario-based uncertainty modeling: DG planning [16]

- Monte Carlo simulation: FACTS allocation [17]
- Robust optimization: Transmission expansion planning [18]
- Fuzzy modeling: distribution network planning [19]

Some power system operation problems are listed as follows:

- Distribution feeder reconfiguration [20]: the on/off statuses of switches are optimally determined to change the network configuration. In distribution networks, it can reduce the active losses by improving the voltage profile and reliability.
- Generation scheduling: UC [21, 22], Dynamic economic dispatch [23], Economic dispatch [24]: the generation level of power plants are determined to minimize the operating costs or maximizing the economic benefits.
- Maintenance scheduling [25]: The maintenance period of assets are determined to keep the adequacy level of the remaining system sufficient while minimizing the costs.
- Optimal power flow: DC OPF [26] and AC OPF [27] are two forms of OPF. In an OPF problem, the decision variables are generation and voltage level of generating units to minimize the operating costs. In DC OPF, it is assumed that all voltage values are 1 pu.
- Active power Loss minimization: minimizing the active losses is a way of improving the efficiency of power system [28].
- Voltage profile improvement [29]: keeping the voltage magnitudes within the normal operating limits is done by changing different decision variables such as reactive power management and demand response.
- Optimal transmission switching [30]: transmission network switching is a technique for changing network topology at the transmission level. It is demonstrated that this can lead to operating cost reduction if it is optimally managed.
- Electric vehicle charging [31]: the charging and discharging of electric vehicles are optimally determined to provide some flexibilities for distribution network operator.
- Offering strategy [32, 33]: the generating units submit their offers to the market operator. The purpose is to maximize the financial benefits.
- Scheduling of reserve [34, 35]: finding the optimal reserve quantity makes the system robust against contingencies and disturbances.
- Loss payments minimization [36]: this approach tries to minimize the payments toward the losses, not the losses by considering the market issues.
- Congestion management [37, 38]: network congestion would reduce the competition level at electricity market. Reducing the congestion would improve the market efficiency.
- Demand side management [39, 40]: the demand side management is harvesting the flexibility from the demand side. The customers are encouraged to shift/reduce their demands to minimize the system requirements for providing services.
- Risk and uncertainty modeling in operation studies:

  - Information gap decision theory (IGDT): Wind operation modeling in OPF [41], unit commitment [42]
  - Scenario-based uncertainty modeling [43]

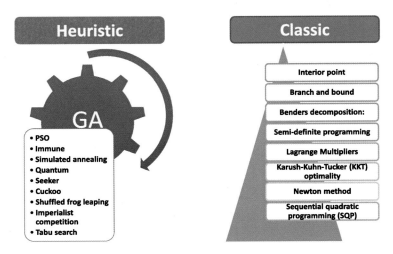

Fig. 1.2 Optimization methods in power system studies

- Monte Carlo simulation: reliability and risk analysis [44]
- Robust optimization: Demand response [40], loss payments minimization [36]
- Fuzzy modeling: DG impact assessment [45]

The methods used for solving the aforementioned optimization problems are categorized into classic and heuristic methods as shown in Fig. 1.2.

Some of the classic methods are listed as follows:

- Interior point: optimal reactive dispatch [46]
- Branch and bound: economic dispatch with disjoint prohibited zones considering network losses [47]
- Benders decomposition: transmission network design problems [48]
- Semi-definite programming: large scale OPF [49, 50]
- Lagrange Multipliers: pricing energy and ancillary services [51]
- Karush-Kuhn-Tucker (KKT) optimality condition: formulation of the terrorist threat problem [52]
- Newton method: Optimal power flow [53, 54]
- Sequential quadratic programming (SQP): UC [55], VAr compensation [56]

Most of the classic methods are gradient-based techniques (in nonlinear problems). This makes them unsuitable for large scale optimization problems. Solving the

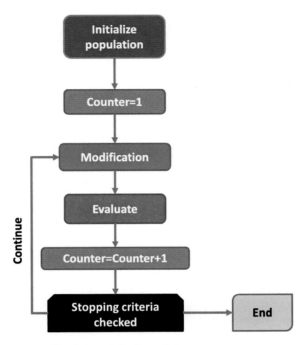

Fig. 1.3 The structure of heuristic optimization techniques

nonlinear problems with integer variables or non-convex constraints would be another challenge for classic techniques.

The heuristic methods are inspired by nature. The basic concept of these techniques is described in Fig. 1.3. In every heuristic method, an initial random population is generated, and it is tried to improve it by using some operator that modify the population (solution). The way each method modifies the population distinguishes that method from the other techniques. For example, the genetic algorithm uses the crossover and mutation operators to improve the solution regarding optimizing the objective function while satisfying the constraints. The particle swarm optimization uses the best solution found by the group and the individual particle to find the optimal solutions. The heuristic methods are also called iterative techniques. The number of iterations needed for optimizing the objective function has an inverse relation with the population number. If the population number is increased, then it can better explore the solution space; however, it takes more time to run. There is always a tradeoff between these two quantities. There are some challenges associated with heuristic techniques. Some of them are listed as follows:

- The parameter tuning is a challenge in these techniques which is usually problem dependent and should be tuned by the decision maker.
- It is not easy to check if the obtained solution is globally optimal or not.

- These methods are usually computationally expensive. In other words, it takes a long time for the decision maker to run. This makes these methods inconvenient for real-time applications.
- Since these methods are iterative, at every iteration, a new solution might be found (which is somehow better than the solution found in previous iterations). Unfortunately, if the problem is solved again (even with the same tuning parameters), it is not guaranteed to reach the similar results.
- Setting the stopping criteria is difficult. This is because the optimal solution of the problem in unknown. When to stop? It is a challenging question to answer. The decision maker usually ends the procedure when there is no significant change in objective function to a maximum number of iterations so reached.
- These solutions are not generally well accepted by the industry.

Some of the heuristic methods applied to power system studies are given as follows:

- Single objective genetic algorithm (GA): network reconfiguration [57]
- Multi-objective Non-dominated sorting genetic algorithm (NSGA): Transmission expansion planning [58]
- Particle swarm optimization (PSO): DG planning [16]
- Immune algorithm (IA): secondary voltage control [59]
- Simulated annealing (SA): maintenance scheduling [60]
- Quantum-inspired evolutionary algorithm: Real and Reactive Power Dispatch [61]
- Seeker Optimization Algorithm: Reactive Power Dispatch [62], coordination of directional over-current relays [63]
- Cuckoo search algorithm: Non-convex economic dispatch [64], capacitor allocation [65]
- Shuffled frog leaping algorithm: unit commitment [66]
- Imperialist competition algorithm: dynamic economic power dispatch [67]
- Tabu search: Economic dispatch [68]
- Ant colony algorithm: Reconfiguration and capacitor placement for loss reduction of distribution systems [69]

1.2 GAMS Installation

The first step is downloading the appropriate installation file from the following address:

https://www.gams.com/download/

The GAMS installation package is available for the following platforms:

Platform	Description
MS Windows 32 bit	Windows Vista or newer on AMD- or Intel-based (x86) architectures.
MS Windows 64 bit	Windows Vista or newer on AMD- or Intel-based (x64) architectures.
Linux 64 bit	AMD- or Intel-based 64-bit (x64) Linux systems with glibc 2.7 or higher.
MacOS X	Intel-based 64-bit (x64) Macintosh system with MacOS X 10.10 (Yosemite) or higher.
Solaris i86pc	AMD- or Intel-based 64-bit (x64) Solaris system. Built on Solaris 11.0.
Solaris SPARC 64bit	Sparc-based 64-bit (sparc-64) Solaris system. Built on Solaris 10.
IBM AIX	PowerPC based 64-bit (ppc-64) AIX system. Built on AIX 6.1

The GAMS interface is depicted in Fig. 1.4.

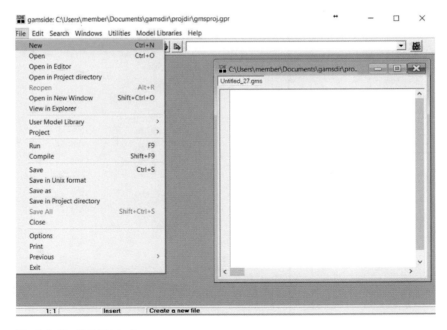

Fig. 1.4 The GAMS interface

1.3 GAMS Elements

Each GAMS model consists of the following main elements:

- Sets: sets are used to define the indices in the algebraic representations of models. For example, set of generating units, set of network buses, set of slack buses, set of time periods, etc.
- Data: The input data of each GAMS model are expressed in the form of Parameters, Tables, or Scalars. The parameters and tables are defined over the sets. The scalars are single value quantities.
- Variables: The variables are decision sets and are unknown before solving the model.
- Equations: The equations describe the relations between the data and variables.
- Model and Solve Statements: The model is defined as a set of equations which contain an objective function. The solve statement asks GAMS to solve the model.
- Output: There are several ways to see the outputs of the solved model such as saving them in XLS file and displaying them.

The General GAMS code structure and elements are shown in Fig. 1.5.

The GAMS elements are explained in a simple example as follows: Suppose a factory produces tree types of products $P_{1,2,3}$. Each product should be processed on two different machines. The available machine hours per day and the time required

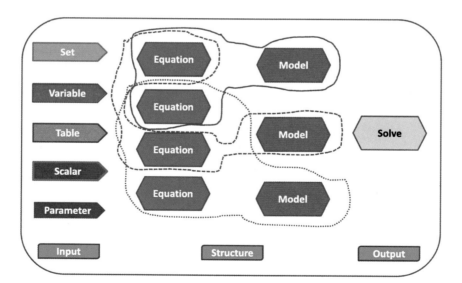

Fig. 1.5 General GAMS code structure and elements

Table 1.1 Data for
illustrative example

Required time for task completion (h)			
Machine	P_1	P_2	P_3
M_1	2	5	2
M_2	3	4	1
Profit per kg ($/kg)		Machine availability (h)	
P_1	10	M_1	16
P_2	12	M_2	12
P_3	13.5		

for each product are considered as the input data. The profits/kg of each item is also
known. The input data of this example is described in Table 1.1.

- Sets: Two sets should be defined: machines (M) and products (P) Sets $M/m1 *$
  $m2/, P/p1 * p3/$;
- Data: Looking at Table 1.1 shows that we have a table (required time for task
  completion) and two parameters (Profit per item and Machine availability). The
  task completion table is defined over two sets (machines and products), the profit
  per item parameter is defined over products while the machine availability is
  defined over product set. It is assumed that the minimum required amount of
  each product is 1 kg.

```
Parameter profit(P)
/p1 10
p2 12
p3 13.5/;
Parameter availability(M)
/m1 16
m2 12/;
Table task(m,p)
p1 p2 p3
m1 2 5 2
m2 3 4 1;
```

- Variables: We are about to decide how many kg should be produced of each
  product. This variable is defined over product set. Another variable should be
  defined as the total profits which should be optimized. This variable (objective
  function) should not have any set index. Variables are OF, $X(p)$.
- Equations: One equation should describe the objective function and another one
  should enforce the constraint for availability of each machine.

```
Equations eq1, eq2;
eq1(m).. sum(p, task(m,p)*x(p))=l=availability(M);
eq2 .. of=e=sum(p, profit(p)*x(p));
```

As it can be seen in this example, there are two types of equations as follows:

- Scalar equation: eq2 is an example of a scalar equation which is not defined over any set.
- Indexed equation: eq1 is an example of the indexed equation. We have to define the equation on index *m* since the index *m* exists in the formulation. In other words, the equation is valid for any element in the set *M*.

Three different relations can be defined in equations as follows:

- =e= Equality: this means that both sides of the equations should be equal to each other.
- =g= Greater than or equal: this means that the left-hand side of the equation should be greater than or equal to the right-hand side of the equation
- =l= Less than or equal: this means that the left-hand side of the equation should be less than or equal to the right-hand side of the equation

- Model: This example has two equations. Both should be included in the model definition.

```
Model example /all/;
```

It can also be defined as follows:

```
Model example /eq1,eq2/;
```

- Solve statement: The solve statement tells GAMS that the model is linear and the direction of the optimization should be toward maximizing the total benefits.

```
Solve example us LP Max OF;
```

The model type used in this code is linear programming (LP). There are various models that can be coded in GAMS coding as follows:

LP: linear programming
QCP: Quadratic programming (the model can only contain linear and quadratic terms)
NLP: Nonlinear problem with continuous constraints
DNLP: Nonlinear problem with discontinuous constraints
MIP: Mixed-integer linear programming
MIQCP: Mixed-integer quadratic constraint programming
MINLP: Mixed-integer nonlinear programming
RMIP: Mixed-integer problem where the integer variables are relaxed

- Output: The outputs of the GAMS model can be displayed and also saved in an XSL file.

```
Display X.l,Of.l;
execute_unload "Example.gdx" X.l
execute 'gdxxrw.exe Example.gdx var=X    rng=Product!a1'
execute_unload "Example.gdx" OF.l
execute 'gdxxrw.exe Example.gdx var=OF    rng=OF!a1'
```

The overall GAMS code for solving the illustrative example is provided in GCode 1.1.

GCode 1.1 Illustrative GAMS example

```
Sets  M /m1*m2/, P /p1*p3/ ;
variables OF,X(p);
parameter profit(P)
/p1  10
 p2  12
 p3  13.5/;
parameter availability (M)
/m1  16
 m2  12/;
Table task(m,p)
     p1      p2      p3
m1    2       5       2
m2    3       4       1  ;
equations eq1,eq2;
eq1(m)..  sum(p, task(m,p)*x(p))=l=availability(M);
eq2    ..  of=l=sum(p,  profit(p)*x(p));
X.lo(p)=1;
model example / all /;
Solve example us LP max OF;
display X.l,Of.l;
execute_unload "Example.gdx" X.l
execute 'gdxxrw.exe Example.gdx var=X   rng=Product!a1'
execute_unload "Example.gdx" OF.l
execute 'gdxxrw.exe Example.gdx var=OF   rng=OF!a1'
```

Fig. 1.6 The structure of
GAMS listing file

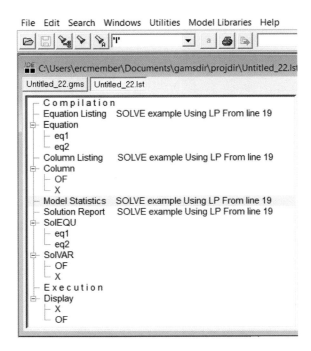

Once the model is solved the solve summary is available by clicking on the file.lst. The structure of GAMS listing file is shown in Fig. 1.6.

The listing file is explained as follows:

S O L V E S U M M A R Y
MODEL example OBJECTIVE OF
TYPE LP DIRECTION MAXIMIZE
SOLVER CPLEX FROM LINE 23
**** SOLVER STATUS 1 Normal Completion
**** MODEL STATUS 1 Optimal
**** OBJECTIVE VALUE 82.7500

The summary shows that the name of the model is *example*. The objective function to be optimized is named OF. The type of the model is linear programming (LP). The direction of the optimization is maximization. The solver used for solving the LP model is CPLEX. The solver status is 1 which means that the model is normally solved without error. The model status is 1 optimal. It means that the global optimal solution is found. The value of the objective function (OF) is 82.75. Different solver status might be reported once the model is solved.

SOLVER STATUS CODE DESCRIPTION
1 Normal Completion
2 Iteration Interrupt
3 Resource Interrupt
4 Terminated by Solver
5 Evaluation Error Limit
6 Capability Problems
7 Licensing Problems
8 User Interrupt
9 Error Setup Failure
10 Error Solver Failure
11 Error Internal Solver Error
12 Solve Processing Skipped
13 Error System Failure

Different model status might be reported once the model is solved.

MODEL STATUS CODE DESCRIPTION
1 Optimal
2 Locally Optimal
3 Unbounded
4 Infeasible
5 Locally Infeasible
6 Intermediate Infeasible
7 Intermediate Nonoptimal
8 Integer Solution
9 Intermediate Non-Integer
10 Integer Infeasible
11 Licensing Problems - No Solution
12 Error Unknown
13 Error No Solution
14 No Solution Returned
15 Solved Unique
16 Solved
17 Solved Singular
18 Unbounded - No Solution
19 Infeasible - No Solution

The model statistic can provide some useful information regarding the developed model. It is indicating that there are two blocks of equations (eq1,eq2) in the developed model. The single equations are three because eq1 has two single equations $(m1, m2)$ and the eq2 has only one equation. There are two blocks of variables x, OF but since X has three variables $(p1, p2, p3)$, then the number of total variables is 4.

> MODEL STATISTICS
> BLOCKS OF EQUATIONS 2 SINGLE EQUATIONS 3
> BLOCKS OF VARIABLES 2 SINGLE VARIABLES 4

By clicking on the SolVar tab, some useful information regarding the variables will be obtained as follows:

> LOWER LEVEL UPPER MARGINAL
> —— VAR OF -INF 82.750 +INF .
> —— VAR X
> LOWER LEVEL UPPER MARGINAL
> p1 1.000 1.000 +INF -3.500
> p2 1.000 1.000 +INF -21.750
> p3 1.000 4.500 +INF .

This means that the minimum limit of variable X is 1 and the upper limits are $+\infty$. The marginal values are also revealed. Each variable in GAMS has five attributes as follows:

- Variable.lo: indicates the minimum limit of a variable
- Variable.up: indicates the maximum limit of a variable
- Variable.l: indicates the level of the variable. In other words, this is the actual value of the variable.
- Variable.fx: indicates the level of the variable is fixed and is not changing. It has the same impact of defining the low and up attributes the same as each other.
- Variable.m: indicates the marginal value of a variable. It shows how much sensitive is the objective function to the changes of this variable. For example, $X.m(p1) = -3.5$ this means that for 1 unit of increase in $X(p1)$ the objective function will reduce by -3.5.

The variables used in this code are all real variables; however, different types of variables can be defined in GAMS as follows:

Variable type	Description	Lower limit	Upper limit
Free	No bounds on variable. Both bounds can be changed from the default values by the user	$-\infty$	$+\infty$
Positive	The positive variable can only take positive values	0	$+\infty$
Negative	The negative variable can only take negative values	$-\infty$	0
Binary	Discrete variable that can only take values of 0 or 1	0	1
Integer	Discrete variable that can only take integer values between the limits	0	100

It should be noted that the default values for variable limits can be modified by the user.

1.4 Conditional Statements

The Dollar ($) condition is broadly used in GAMS coding. Various applications are explained through some examples.

- Suppose we need to have a conditional statement like this if $A = C$ then $B = D$. This statement is modeled as follows:

(B=D)$(A=C)

If $(A \neq C)$ then no assignment to B will happen.

- If $A = C$ then $B = D$ otherwise $B = 0$. This statement is modeled as follows:

B=D$(A=C)

- Suppose that we need to filter some elements in eq1(m). In other words, this equation should be valid for every m elements except some of them.

eq1(m)$(ord(m)=1) .. sum(p, task(m,p)*x(p))=l=availability(M);

This would force the equation only for $m1$.

```
eq1(m)$(ord(m) ≥ 2) .. sum(p, task(m,p)*x(p))=l=availability(M);
```

This would force the equation only for $m2$ and $m3$. It should be noted that no variable can exist in conditional statement.

- If A and B then $C = D$, this is coded as follows:

```
(C=D)$( A and B)
```

1.5 Loop Definition in GAMS

The looping is used for executing one or more statements repeatedly. There are different forms of looping in GAMS as follows:

1.5.1 LOOP Statement

One of the techniques for defining the loop in GAMS is loop statement.

```
Loop(setname,
Statement 1 ;
Statement 2 ;
Statement 3 ;
);
```

The statements 1–3 are executed N times. N is equal to the cardinality of the "setname." These statements can assign values to the variable limits or describe the relation between some parameters (not the variables). No variable can appear in the loop statements.

```
set counter /c1*c4/;
Loop(counter,
X.lo(p)=0.1*ord(counter);
Solve example us LP max OF;
);
```

In the first counter, the minimum values of all X variables would be $0.1*1$ and the model is solved. The second counter sets the minimum values of X variable equal to $0.1*2 = 0.2$ and then the model is solved. The point that should be noted here is that the set/parameter/variable/model definition should be outside the loop.

This kind of modeling would be useful in conducting sensitivity analysis. Some parameters are varied and then the impacts on the objective function are investigated as follows:

```
set counter /c1*c4/;
parameter report(counter,*);
Loop(counter,
X.lo(p)=0.1*ord(counter);
Solve example us LP max OF;
report(counter,'lowerlimit')=0.1*ord(counter);
report(counter,'OF')=OF.l;
);
```

1.5.2   WHILE Statement

One of the techniques for defining the loop in GAMS is while statement.

```
while (condition,
Statement 1 ;
Statement 2 ;
Statement 3 ;
);
```

The statements 1–3 are executed as far as the condition is true. The condition will be changed inside the loop based on some logic otherwise the loop will continue forever!

1.5.3   FOR Statement

One of the techniques for defining the loop in GAMS is FOR statement.

```
scalar Itermax /10/;
scalar iteration ;
for (iteration=1 to Itermax,
Statement 1 ;
Statement 2 ;
Statement 3 ;
);
```

The statements 1–3 are executed for Itermax times.

1.5.4   REPEAT-UNTIL Statement

One of the techniques for defining the loop in GAMS is Repeat-until statement.

```
repeat ( Statement 1 ;
Statement 2 ;
Statement 3 ;
until condition );
```

The statements 1–3 are executed until the condition becomes logically true (it should be initially false).

1.6   Linking GAMS and Excels

1.6.1   Reading from Excel

GAMS is able to read the data from xls files. It is also convenient to use xls as an interface between GAMS and other platforms. Reading the data from xls file is straightforward as follows:

```
Parameter Level(m,p);
\$CALL GDXXRW.EXE Example.xls par=Level rng=task!E5:H7
\$GDXIN Example.gdx
\$LOAD Level
```

In this case, a parameter like Level (m, p) is defined. The xls file name (Example.xls) is indicated and the range of data is specified (sheet=task, cells:

E5:H7). The data is read from xls file and is written on Example.gdx file. The parameter level is loaded and can be used in the code.

1.6.2   Writing to Excel

It is usually desirable to save the output of a GAMS code in an excel file. The procedure is quite straightforward.

For variables, the following commands should be executed. The user can indicate which variable (X.l,OF.l) should be written in Excel file. The name of the excel file (Example.xls) and what would be the sheet name (product, OF). The cell a1 is chosen as the starting cell in the specified sheet for writing the data.

```
execute_unload  "Example.gdx"  X.l
execute  'gdxxrw.exe  Example.gdx  var=X   rng=Product!a1'
execute_unload  "Example.gdx"  OF.l
execute  'gdxxrw.exe  Example.gdx  var=OF   rng=OF!a1'
```

If the user wants to write the parameter X to an excel file then the command would be as follows:

```
execute_unload  "Example.gdx"  X
execute  'gdxxrw.exe  Example.gdx  Par=X   rng=Product!a1'
```

In this case, no '.l' is needed since it is a parameter, not a variable.

1.7   Error Debugging in GAMS

Once the code is written the "Shift key + F9" should be pressed. The GAMS compiler will check the developed code without executing it. The coding errors are unavoidable in every programming language, and GAMS is not an exception. Usually, a list of errors is generated which should be taken care of. The first thing to keep in mind is to start from the first error and fix them one by one. In this section, some common coding errors and best way of debugging them are explained.

- Error 140: Unknown symbol: GAMS compiler generates this error when there are some undefined symbols in the model.
- Error 246: Objective variable is not a free variable

    This error will usually happen when the objective function (which should be minimized/maximized) is not set as a free variable. For example, it is defined as a positive/negative variable. In case that the decision maker needs the objective function be a positive variable, the lower limit of the variable should be specified using *.lo* = 0 command.

- Error: Unbounded variable:

  This means that the objective function is equal to $-\infty$ or $+\infty$. This usually happens when the bounds of the objective function are not properly defined, or optimization direction is not properly defined in solve statement. For example, if the objective is to be minimized but it is maximized by mistake. The following error will be generated by GAMS.

**** ERRORS/WARNINGS IN VARIABLE OF
1 error(s): Unbounded variable

- Error 149: Uncontrolled set entered as constant

  This error usually happens when the statements are not properly written. Consider the following line of code

```
eq1 .. sum(p, task(m,p)*x(p))=l=availability(M);
```

The equation should be valid for every m, but the equation label is not defined over the set m. The correct format for defining this equation is as follows:

```
eq1(m) .. sum(p, task(m,p)*x(p))=l=availability(M);
```

- Error 143: A suffix is missing

  This error will happen when a variable is used outside of the equation environment. For example, the following line is trying to assign a value to variable $X(p)$ but not in any equation:

```
x(p)=1;
```

This is not a valid way of doing this. If the variable $X(p)$ should be fixed to a constant value it should be defined as follows:

```
x.fx(p)=1;
```

As a general rule, the variables cannot appear outside the equations. Only the variable's attributes can be modified outside the equation environment.

This means $X.lo(p)$, $X.l(p)$, $X.up(p)$ can be changed without using equation environment.

Another situation that this error might happen is in display command. Consider the following part of a GAMS code:

```
Solve example us LP max Z; Display X , Of.l;
```

This will generate the following error:

```
143 A suffix is missing
**** 1 ERROR(S) 0 WARNING(S)
```

This is because X is a variable and display command can only show the attributes of the X variable. The correct form for showing the value of X variable is as follows:

```
solve example us LP max Z;
Display X.l , Of.l;
```

- Error 245: Objective variable not referenced in model
  This means that the solve statement is trying to minimize/maximize a variable which does not exist in the defined model.
- Error 141: Symbol declared, but no values have been assigned
  This error happens when the variable or parameter is defined, but it is not implicitly assigned any value and also does not appear in any equation. In other words, GAMS has no info about this symbol, but we are asking GAMS some information about it. Consider the following GAMS code:

```
Sets M /m1*m2/, P /p1*p3/ ;
Variables OF,x(p),Z;
parameter profit(P)
/p1 10
p2 12
p3 13.5/;
Parameter availability (M)
/m1 16
```

(continued)

```
m2 12/;
Table task(m,p)
p1 p2 p3
m1 2 5 2
m2 3 4 1 ;
Equations eq1,eq2;
eq1(m) .. sum(p, task(m,p)*x(p))=l=availability(M);
eq2 .. of=l=sum(p, profit(p)*x(p));
X.lo(p)=1;
Model example /all/;
Solve example us LP max OF;
display X.l , Of.l , Z.l;
```

This code would generate Error 141 because although variable Z is defined but since it has no initial assignment and does not appear in the model then GAMS knows nothing about it.

- Error 225: Floating entry ignored

  This error is a very common error specially for beginners. This happens when a table is not properly typed in GAMS. All data should be aligned under the column label.

	P1	P2	P3
M1	2	5		2
M2	3	4	1

The correct form of typing the above table is as follows:

	P1	P2	P3
M1	2	5	2
M2	3	4	1

Handling the large tables in GAMS would be much easier if the tool named "xls2gams.exe" is used. This tool is available in the same folder that GAMS is installed. The function of this tool is copying data from xls file into GAMS format. Using this tool to import xls data to GAMS is highly recommended.

- Error 56: Endogenous operands for * not allowed in linear models

Consider the following code:

```
Sets M /m1*m2/, P /p1*p3/ ;
Variables OF,x(p),Z;
parameter profit(P)
/p1 10
p2 12
p3 13.5/;
Parameter availability (M)
/m1 16
m2 12/;
Table task(m,p)
p1 p2 p3
m1 2 5 2
m2 3 4 1 ;
Equations eq1,eq2;
eq1(m) .. sum(p, task(m,p)*x(p))=l=availability(M);
eq2 .. of=l=sum(p, profit(p)*x(p)*x(p));
X.lo(p)=1;
Model example /all/;
Solve example us LP max OF;
display X.l , Of.l , Z.l;
```

The following error will be generated after compiling this code. In order to find out why this error is happening we should double-check the model. The solve statement is indicating that the model is LP but in eq2 we can see $profit(p) * x(p) * x(p)$. This means that the model is not LP and solve statement should be modified as follows:

```
Solve example us NLP max OF;
or
Solve example us QCP max OF;
```

- Error 37:
  Consider the following line in a GAMS code:

```
eq2 .. of=sum(p, profit(p)*x(p));
```

This line is not properly defined because the correct operator for stating the equality is not used in equation environment. The only operators that can be used are '=l=' or '=e=' or '=g='. the correct format is as follows:

```
eq2 .. of=e=sum(p, profit(p)*x(p));
```

- Error 409:
  Consider the following lines in a GAMS code:

```
eq1(m) .. sum(p, task(m,p)*x(p))=l=availability(M)
eq2 .. of=sum(p, profit(p)*x(p)*x(p));
```

This is because a semicolon (;) is missing at the end of eq1.

1.8 General Programming Remarks

Like other programming languages, there are different ways of modeling the problem in GAMS. The so-called "best way" is yet to be discovered. Some programming tips are given in this section as follows:

- GAMS compiler is not case sensitive. The lowercase letters and capital letters are not distinguished. For example, $P(t)$ is the same as $p(T)$ or $P(T)$ or $p(t)$.
- Try to add explanations for your equations and symbols you define in your code. Use * for commenting a line or $ontext $offtext for a block of comments.
- Choose the symbol names carefully and meaningfully. For example for the power generated by power plants, $P(g)$ is a better choice than $X(i)$. You can define g as the set of generating units and variable P as the generated power.
- Don't get frightened by the errors once your code is compiled. The best thing to see is the error flag. Sometimes although the model does not generate any error flag (since you have obeyed the GAMS syntax) but it is not correctly modeling the given problem.
- Keep in mind that the machine is always right so GAMS cannot go wrong. Debugging is also a part of the coding process.
- Always check the GAMS output. The variables, model statistics, and solver statistics should be checked to see if any flag is raised. If the output of GAMS model is not what you expect then there should be something wrong with your model (not the GAMS). Double-check the equations and see if the developed model is actually what it is desired to be or not.

- If you need to specify some limits for your variables then try to do it using .lo or .up attributes. Although creating a new equation for modeling this constraint is "legal", however, it is totally "inappropriate". For example, you need to state that the minimum value of variable $P(g)$ is 1 then it can be easily done using $P.\text{lo}(g) = 1$. Sometimes the variables should be limited using some tables of parameters. For example, the minimum values of $P(g)$ are stored in a parameter $P\min(g)$ then $P.\text{lo}(g) = P\min(g)$ will do the job. As far as no variable exists then no equation is needed. If $P\min(g)$ is a variable, then the following line will be appropriate for modeling this constraint.

```
eq(g) .. P(g)=g=Pmin(g);
```

It should be noted that *eq* is defined over the set *g* since the inequality should be valid for every element belonging to the set *g*.
- Although GAMS is a robust programming language less number of variables and equations are always welcome. Try to avoid unnecessary equations and variables. Use filtering to omit unnecessary equations. For example, we need to define an equation which calculates the power flow of a line connecting 'bus' to 'node' in DC power flow formulation. The following GAMS code might be the first attempt to model it.

```
Eq(bus,node) .. Pij(bus,node)=e=(d(bus)-d(node))/data(bus,node,'x');
```

Suppose the network has 186 lines and 118 buses. This kind of coding would create two problems:

- The parameter data(bus, node,'x') is zero when bus and node are not connected to each other. GAMS might generate a division by Zero error! It's not always easy to find and remove the cause of error.
- If you check the model you can see that this equation adds 13,924 single equations to the model. You might ask yourself why? We were expecting to have only 186 equations (or double that since flow in opposite direction should also be calculated). It is almost 74 times bigger than what we were expecting to have. This is because GAMS is considering every combination between "bus" and "node" (118*118).

The better way to code this line is as follows:

```
Eq(bus,node)$branch(bus,node,'x')..Pij(bus,node)=e=(d(bus)-
d(node))/data(bus,node,'x');
```

In this way, the equation is filtered and is only valid for those lines which have nonzero reactance values. It also reduces the computation time and the chance of difficulties in finding a feasible/optimal solution.

- Don't forget to put a semicolon (;) at the end of every line (sometimes it is not necessary but if you have just started coding in GAMS, do it).

- Different solvers might use different approaches to solve the model. Try different solver to find out which one is more successful/faster in finding a solution for your problem.

- The default value for undefined elements of a table or parameter is zero. For example, consider a parameter like Data(bus,node) representing the line reactance; Suppose that we define Data('2','4')=0.5; If this parameter is displayed, you can see that all elements are zero except Data('2','4'). If you are a power engineer, you know that the reactance between bus number 2 and bus number 4 should be the same as reactance between bus number 4 and bus number 2. In other words, it should be symmetrical. Unfortunately (or fortunately) GAMS is not a power system engineer and does not understand even a word from power system or electrical engineering. GAMS only understands whatever it is told by the user (code developer).

- Go to your GAMS model library and have a look at those developed codes. There are loads of new things to learn even for experienced GAMS code developers. The GAMS model library can be accessed through the interface as shown in Fig. 1.7.

- Try to provide all of the variables appropriate upper and lower limits as well as initial starting values. The initial values can be assigned to the variables using .l expression (do it before solving statement). For example, voltage variables can be assigned 1 per unit values as the initial starting values. This is a double-edged sword. This is because providing a good starting value can highly improve the solution procedure especially in nonlinear or mixed integer nonlinear models but a poor starting value might slowdown the solver.

- It is very likely that you get an infeasible status for your model. Especially when the model is large and includes a large number of variables. The following steps might be helpful in resolving the problem:

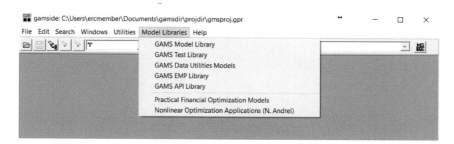

Fig. 1.7 GAMS model library

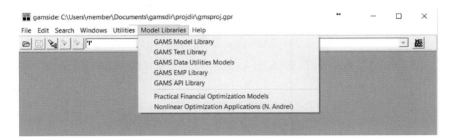

Fig. 1.8 GAMS help menu

- Check to see if you can better express your model. It is always desired to stay as close as possible to the linear form of expressing the equations.
- Provide better initial values for your variables.
- Relax the variable limits, rerun the model. If the problem is resolved, then it means that the variable limits should be revised. If your model contains integer/binary variables, then you can solve the model using relaxed option. For example, the MIP model can be solved using RMIP. This would ask the GAMS to neglect integer nature of the variables (the variable limits remain unchanged). For relaxing the MINLP and MIQCP, you should use RMINLP and RMIQCP, respectively.
- Remove some equations or add some slack variables to the model to see if you can find the trouble making equations
- Ask support from those experienced GAMS code developers (if they have time and are willing to contribute to your project). This option is intentionally placed at the end of suggestion list.

There is a set of good resources that can be accessed through the GAMS interface as it is shown in Fig. 1.8. Additionally, some other useful tutorials are listed as follows:

- A GAMS tutorial [70]
- GAMS language guide [71]
- GAMS—Modeling and Solving Optimization Problems [72]
- GAMS and MATLAB interface [73–75]
- Grid-enabled optimization with GAMS [76]
- Practical financial optimization in GAMS [77]

1.9   Book Structure

The objectives pursued in this book are the following:

- Familiarizing the readers with optimization concept through multiple examples
- Introducing different optimization problems that the decision makers might face in power system studies

- Providing robust solutions for different operation/planning problems in power system studies
- Exploring the capabilities of GAMS in conducting sensitivity analysis

This book consists of ten chapters. A brief description of each chapter is provided here:

- Chapter 1 provides the instructions on how to start coding in GAMS. Different programming elements and some coding tricks are introduced and explained. Reading this chapter would be helpful to understand what to expect from GAMS in dealing with optimization problems.
- Chapter 2 provides some insights to the reader on how to solve simple optimization problems through some illustrative examples. These examples are pure mathematics, and no power system or electrical engineering background is required for the readers. This chapter can also be used for those who might be only interested in learning GAMS for interdisciplinary applications.
- Chapter 3 discusses how to model the dispatching problem of different power plants in a single snapshot (single period). Different power plant technologies are explained such as thermal power, wind turbine, CHP, and hydro-power plants.
- Chapter 4 provides a solution for dynamic economic dispatch problem (multi-period problems). The time-dependent optimal dispatch decisions are modeled and solved.
- Chapter 5 explains how unit commitment problem can be modeled and solved in GAMS. The developed codes in this chapter are linear and categorized as mixed integer linear programming (MIP) models. The inputs are generator's characteristics, electricity prices, and demands. The outputs of these codes are on/off status of units and their operating schedules.
- Chapter 6 provides a solution for optimal power flow (OPF) problem. The active/reactive power output of generating units as well as the network variables (voltage magnitudes and angles) are determined in OPF to minimize total operating costs. Different OPF models are investigated such as single and multi-period DC/AC optimal power flow. The appropriate linear/nonlinear models are developed and solved in this chapter.
- Chapter 7 provides a solution for modeling the operation and planning problems of energy storage systems (ESS). The inputs are generator's characteristics, electricity prices, demands, and network topology. The outputs of this code are operating schedules/locations and sizes of ESS units.
- Chapter 8 provides a solution for allocation of Phasor Measurement Units (PMU) problem in transmission networks to maximize the power system observability. The PMU can measure the voltage phasor at the connection bus, and also it measures the current phasor of any branch connected to the bus hosting the PMU. Different cases are analyzed, and the problem is tested on some standard IEEE cases.
- Chapter 9 provides a solution for some transmission network operation and planning studies in GAMS. The transmission investment regarding building new lines and power flow controllers (phase shifter), sensitivity factors, and

transmission switching have been discussed in this chapter. The GAMS code for solving each optimization problem is developed and discussed.

- Chapter 10 provides a solution for Energy System Integration (ESI) problem in GAMS. The question which is answered in this chapter is how to harvest the flexibilities in different energy sectors by coordinated operation of these sectors.

References

1. J.L.C. Meza, M.B. Yildirim, A.S.M. Masud, A model for the multiperiod multiobjective power generation expansion problem. IEEE Trans. Power Syst. **22**(2), 871–878 (2007)
2. S. Kannan, S. Baskar, J.D. McCalley, P. Murugan, Application of NSGA-II algorithm to generation expansion planning. IEEE Trans. Power Syst. **24**(1), 454–461 (2009)
3. R. Fang, D.J. Hill, A new strategy for transmission expansion in competitive electricity markets. IEEE Trans. Power Syst. **18**(1), 374–380 (2003)
4. R. Romero, A. Monticelli, A. Garcia, S. Haffner, Test systems and mathematical models for transmission network expansion planning. IEE Proc. Gener. Transm. Distrib. **149**(1), 27–36 (2002)
5. V. Miranda, J.V. Ranito, L.M. Proenca, Genetic algorithms in optimal multistage distribution network planning. IEEE Trans. Power Syst. **9**(4), 1927–1933 (1994)
6. S. Gerbex, R. Cherkaoui, A.J. Germond, Optimal location of multi-type facts devices in a power system by means of genetic algorithms. IEEE Trans. Power Syst. **16**(3), 537–544 (2001)
7. C. Wang, M.H. Nehrir, Analytical approaches for optimal placement of distributed generation sources in power systems. IEEE Trans. Power Syst. **19**(4), 2068–2076 (2004)
8. W. El-Khattam, K. Bhattacharya, Y. Hegazy, M.M.A. Salama, Optimal investment planning for distributed generation in a competitive electricity market. IEEE Trans. Power Syst. **19**(3), 1674–1684 (2004)
9. A. Keane, M. O'Malley, Optimal allocation of embedded generation on distribution networks. IEEE Trans. Power Syst. **20**(3), 1640–1646 (2005)
10. S. Sundhararajan, A. Pahwa, Optimal selection of capacitors for radial distribution systems using a genetic algorithm. IEEE Trans. Power Syst. **9**(3), 1499–1507 (1994)
11. B. Milosevic, M. Begovic, Nondominated sorting genetic algorithm for optimal phasor measurement placement. IEEE Trans. Power Syst. **18**(1), 69–75 (2003)
12. B. Gou, Generalized integer linear programming formulation for optimal PMU placement. IEEE Trans. Power Syst. **23**(3), 1099–1104 (2008)
13. Y.M. Atwa, E.F. El-Saadany, Optimal allocation of ESS in distribution systems with a high penetration of wind energy. IEEE Trans. Power Syst. **25**(4), 1815–1822 (2010)
14. H. Oh, Optimal planning to include storage devices in power systems. IEEE Trans. Power Syst. **26**(3), 1118–1128 (2011)
15. P. Maghouli, A. Soroudi, A. Keane, Robust computational framework for mid-term techno-economical assessment of energy storage. IET Gener. Transm. Distrib. **10**(3), 822–831 (2016)
16. A. Soroudi, M. Afrasiab, Binary PSO-based dynamic multi-objective model for distributed generation planning under uncertainty. IET Renew. Power Gener. **6**(2), 67–78 (2012)
17. G. Blanco, F. Olsina, F. Garces, C. Rehtanz, Real option valuation of facts investments based on the least square monte carlo method. IEEE Trans. Power Syst. **26**(3), 1389–1398 (2011)
18. B. Chen, J. Wang, L. Wang, Y. He, Z. Wang, Robust optimization for transmission expansion planning: minimax cost vs. minimax regret. IEEE Trans. Power Syst. **29**(6), 3069–3077 (2014)
19. V. Miranda, M.A.C.C. Matos, Distribution system planning with fuzzy models and techniques, in *10th International Conference on Electricity Distribution, 1989 (CIRED 1989)*, vol. 6, Brighton (1989), pp. 472–476

20. S. Civanlar, J.J. Grainger, H. Yin, S.S.H. Lee, Distribution feeder reconfiguration for loss reduction. IEEE Trans. Power Delivery 3(3), 1217–1223 (1988)
21. M. Carrion, J.M. Arroyo, A computationally efficient mixed-integer linear formulation for the thermal unit commitment problem. IEEE Trans. Power Syst. 21(3), 1371–1378 (2006)
22. J.M. Arroyo, A.J. Conejo, Optimal response of a thermal unit to an electricity spot market. IEEE Trans. Power Syst. 15(3), 1098–1104 (2000)
23. D.W. Ross, S. Kim, Dynamic economic dispatch of generation. IEEE Trans. Power Apparatus Syst. PAS-99(6), 2060–2068 (1980)
24. Z.-L. Gaing, Particle swarm optimization to solving the economic dispatch considering the generator constraints. IEEE Trans. Power Syst. 18(3), 1187–1195 (2003)
25. J. Endrenyi, S. Aboresheid, R.N. Allan, G.J. Anders, S. Asgarpoor, R. Billinton, N. Chowdhury, E.N. Dialynas, M. Fipper, R.H. Fletcher, C. Grigg, J. McCalley, S. Meliopoulos, T.C. Mielnik, P. Nitu, N. Rau, N.D. Reppen, L. Salvaderi, A. Schneider, C. Singh, The present status of maintenance strategies and the impact of maintenance on reliability. IEEE Trans. Power Syst. 16(4), 638–646 (2001)
26. A.G. Bakirtzis, P.N. Biskas, A decentralized solution to the DC-OPF of interconnected power systems. IEEE Trans. Power Syst. 18(3), 1007–1013 (2003)
27. R.D. Zimmerman, C.E. Murillo-Sanchez, R.J. Thomas, Matpower: steady-state operations, planning, and analysis tools for power systems research and education. IEEE Trans. Power Syst. 26(1), 12–19 (2011)
28. Y.M. Atwa, E.F. El-Saadany, M.M.A. Salama, R. Seethapathy, Optimal renewable resources mix for distribution system energy loss minimization. IEEE Trans. Power Syst. 25(1), 360–370 (2010)
29. K.R.C. Mamandur, R.D. Chenoweth, Optimal control of reactive power flow for improvements in voltage profiles and for real power loss minimization. IEEE Trans. Power Apparatus Syst. PAS-100(7), 3185–3194 (1981)
30. Y. Bai, H. Zhong, Q. Xia, C. Kang, A two-level approach to ac optimal transmission switching with an accelerating technique. IEEE Trans. Power Syst. 32(2), 1616–1625 (2017)
31. N. Rotering, M. Ilic, Optimal charge control of plug-in hybrid electric vehicles in deregulated electricity markets. IEEE Trans. Power Syst. 26(3), 1021–1029 (2011)
32. F. Wen, A.K. David, Optimal bidding strategies and modeling of imperfect information among competitive generators. IEEE Trans. Power Syst. 16(1), 15–21 (2001)
33. A. Baillo, M. Ventosa, M. Rivier, A. Ramos, Optimal offering strategies for generation companies operating in electricity spot markets. IEEE Trans. Power Syst. 19(2), 745–753 (2004)
34. H.B. Gooi, D.P. Mendes, K.R.W. Bell, D.S. Kirschen, Optimal scheduling of spinning reserve. IEEE Trans. Power Syst. 14(4), 1485–1492 (1999)
35. K.A. Papadogiannis, N.D. Hatziargyriou, Optimal allocation of primary reserve services in energy markets. IEEE Trans. Power Syst. 19(1), 652–659 (2004)
36. A. Soroudi, P. Siano, A. Keane, Optimal DR and ESS scheduling for distribution losses payments minimization under electricity price uncertainty. IEEE Trans. Smart Grid 7(1), 261–272 (2016)
37. S. Dutta, S.P. Singh, Optimal rescheduling of generators for congestion management based on particle swarm optimization. IEEE Trans. Power Syst. 23(4), 1560–1569 (2008)
38. C. Murphy, A. Soroudi, A. Keane, Information gap decision theory-based congestion and voltage management in the presence of uncertain wind power. IEEE Trans. Sust. Energy 7(2), 841–849 (2016)
39. B. Hayes, I. Hernando-Gil, A. Collin, G. Harrison, S. Djoki, Optimal power flow for maximizing network benefits from demand-side management. IEEE Trans. Power Syst. 29(4), 1739–1747 (2014)
40. A.J. Conejo, J.M. Morales, L. Baringo, Real-time demand response model. IEEE Trans. Smart Grid 1(3), 236–242 (2010)
41. A. Rabiee, A. Soroudi, A. Keane, Information gap decision theory based OPF with HVDC connected wind farms. IEEE Trans. Power Syst. 30(6), 3396–3406 (2015)

42. A. Soroudi, A. Rabiee, A. Keane, Information gap decision theory approach to deal with wind power uncertainty in unit commitment. Electr. Power Syst. Res. **145**, 137–148 (2017)
43. A.J. Conejo, M. Carrión, J.M. Morales, *Decision Making Under Uncertainty in Electricity Markets*, vol. 1 (Springer, New York, 2010)
44. E. Zio, *The Monte Carlo Simulation Method for System Reliability and Risk Analysis* (Springer, London, 2013)
45. A. Soroudi, Possibilistic-scenario model for DG impact assessment on distribution networks in an uncertain environment. IEEE Trans. Power Syst. **27**(3), 1283–1293 (2012)
46. S. Granville, Optimal reactive dispatch through interior point methods. IEEE Trans. Power Syst. **9**(1), 136–146 (1994)
47. T. Ding, R. Bo, F. Li, H. Sun, A bi-level branch and bound method for economic dispatch with disjoint prohibited zones considering network losses. IEEE Trans. Power Syst. **30**(6), 2841–2855 (2015)
48. S. Binato, M.V.F. Pereira, S. Granville, A new benders decomposition approach to solve power transmission network design problems. IEEE Trans. Power Syst. **16**(2), 235–240 (2001)
49. D.K. Molzahn, J.T. Holzer, B.C. Lesieutre, C.L. DeMarco, Implementation of a large-scale optimal power flow solver based on semidefinite programming. IEEE Trans. Power Syst. **28**(4), 3987–3998 (2013)
50. R.A. Jabr, Exploiting sparsity in SDP relaxations of the OPF problem. IEEE Trans. Power Syst. **27**(2), 1138–1139 (2012)
51. T. Wu, M. Rothleder, Z. Alaywan, A.D. Papalexopoulos, Pricing energy and ancillary services in integrated market systems by an optimal power flow. IEEE Trans. Power Syst. **19**(1), 339–347 (2004)
52. J.M. Arroyo, F.D. Galiana, On the solution of the bilevel programming formulation of the terrorist threat problem. IEEE Trans. Power Syst. **20**(2), 789–797 (2005)
53. D.I. Sun, B. Ashley, B. Brewer, A. Hughes, W.F. Tinney, Optimal power flow by Newton approach. IEEE Trans. Power Apparatus Syst. **PAS-103**(10), 2864–2880 (1984)
54. F. Milano, Continuous Newton's method for power flow analysis. IEEE Trans. Power Syst. **24**(1), 50–57 (2009)
55. E.C. Finardi, E.L. da Silva, Solving the hydro unit commitment problem via dual decomposition and sequential quadratic programming. IEEE Trans. Power Syst. **21**(2), 835–844 (2006)
56. I.P. Abril, J.A.G. Quintero, Var compensation by sequential quadratic programming. IEEE Trans. Power Syst. **18**(1), 36–41 (2003)
57. B. Enacheanu, B. Raison, R. Caire, O. Devaux, W. Bienia, N. HadjSaid, Radial network reconfiguration using genetic algorithm based on the matroid theory. IEEE Trans. Power Syst. **23**(1), 186–195 (2008)
58. P. Maghouli, S.H. Hosseini, M.O. Buygi, M. Shahidehpour, A scenario-based multi-objective model for multi-stage transmission expansion planning. IEEE Trans. Power Syst. **26**(1), 470–478 (2011)
59. T. Amraee, A.M Ranjbar, R. Feuillet. Immune-based selection of pilot nodes for secondary voltage control. Eur. Trans. Electr. Power **20**(7), 938–951 (2010)
60. T. Satoh, K. Nara, Maintenance scheduling by using simulated annealing method [for power plants]. IEEE Trans. Power Syst. **6**(2), 850–857 (1991)
61. J.G. Vlachogiannis, K.Y. Lee. Quantum-inspired evolutionary algorithm for real and reactive power dispatch. IEEE Trans. Power Syst. **23**(4), 1627–1636 (2008)
62. C. Dai, W. Chen, Y. Zhu, X. Zhang, Seeker optimization algorithm for optimal reactive power dispatch. IEEE Trans. Power Syst. **24**(3), 1218–1231 (2009)
63. T. Amraee, Coordination of directional overcurrent relays using seeker algorithm. IEEE Trans. Power Delivery **27**(3), 1415–1422 (2012)
64. D.N. Vo, P. Schegner, W. Ongsakul, Cuckoo search algorithm for non-convex economic dispatch. IET Gener. Transm. Distrib. **7**(6), 645–654 (2013)
65. A.A. El-fergany, A.Y. Abdelaziz, Capacitor allocations in radial distribution networks using cuckoo search algorithm. IET Gener. Transm. Distrib. **8**(2), 223–232 (2014)

66. M. Barati, M.M. Farsangi, Solving unit commitment problem by a binary shuffled frog leaping algorithm. IET Gener. Transm. Distrib. **8**(6), 1050–1060 (2014)
67. R. Roche, L. Idoumghar, B. Blunier, A. Miraoui, Imperialist competitive algorithm for dynamic optimization of economic dispatch in power systems, in *International Conference on Artificial Evolution (Evolution Artificielle)* (Springer, Berlin, 2011), pp. 217–228
68. W.-M. Lin, F.-S. Cheng, M.-T. Tsay, An improved tabu search for economic dispatch with multiple minima. IEEE Trans. Power Syst. **17**(1), 108–112 (2002)
69. C.F. Chang, Reconfiguration and capacitor placement for loss reduction of distribution systems by ant colony search algorithm. IEEE Trans. Power Syst. **23**(4), 1747–1755 (2008)
70. R.E. Rosenthal, A GAMS tutorial. Technical note (1992)
71. A. Brooke, D. Kendrick, A. Meeraus, R. Raman, R.E. Rosenthal, *Gams. A Users Guide* (GAMS Development Corporation, Washington, DC, 2005)
72. A. Geletu, *Gams-Modeling and Solving Optimization Problems* (TU-Ilmenau, Faculty of Mathematics and Natural Sciences, Department of Operation Research & Stochastrics, Ilmenau, 2008)
73. M.C. Ferris, Matlab and GAMS: interfacing optimization and visualization software. Mathematical Programming Technical Report, 98:19 (1998)
74. L. Wong et al., *Linking Matlab and Gams: A Supplement* (University of Victoria, Department of Economics, Victoria, BC, 2009)
75. M.C. Ferris, R. Jain, S. Dirkse, Gdxmrw: Interfacing GAMS and matlab (2011). http://www.gams.com/dd/docs/tools/gdxmrw.pdf
76. M.R. Bussieck, M.C. Ferris, A. Meeraus, Grid-enabled optimization with GAMS. INFORMS J. Comput. **21**(3), 349–362 (2009)
77. S.S. Nielson, A. Consiglio, *Practical Financial Optimization: A Library of GAMS Models* (Wiley, New York, 2010)

Chapter 2
Simple Examples in GAMS

The main concept that is developed in this chapter is explaining some optimization categories that can be modeled in GAMS. These models include linear programming (LP), mixed integer programming (MIP), nonlinear programming (NLP), quadratic programming (QCP), mixed integer non-linear programming (MINLP), and multi-objective optimization problems.

Understanding the materials presented and discussed in this chapter does not require any background in power system studies. This makes it suitable for anybody who might be interested to start optimization modeling in GAMS.

2.1 Different Types of Optimization Models

The general form of an optimization problem is as follows:

$$\min_{X} f(X, I) \tag{2.1a}$$

$$G(X, I) \leq 0 \tag{2.1b}$$

$$H(X, I) = 0 \tag{2.1c}$$

where f is objective function, G and H are set of equality and inequality constraints, respectively, I is the input data of the optimization problem, and X is the set of decision variables that should not only satisfy G and H but also optimizes the f value.

© Springer International Publishing AG 2017
A. Soroudi, *Power System Optimization Modeling in GAMS*,
DOI 10.1007/978-3-319-62350-4_2

2.1.1 Linear Programming (LP)

The linear programming problems are those that f, G, H are all linear in (2.1).

2.1.1.1 LP Example

A simple linear programming example is as follows:

$$\min_{X} \text{OF} = x_1 + 3x_2 + 3x_3 \tag{2.2a}$$

$$x_1 + 2x_2 \geq 3 \tag{2.2b}$$

$$x_3 + x_2 \geq 5 \tag{2.2c}$$

$$x_1 + x_3 = 4 \tag{2.2d}$$

GCode 2.1 LP Example (2.2)

```
variables  x1 , x2 , x3 , of ;
Equations
eq1
eq2
eq3
eq4 ;
eq1  ..  x1+2*x2  =g=3;
eq2  ..  x3+x2  =g=5;
eq3  ..  x1+x3  =e=4;
eq4  ..  x1+3*x2  +3*x3=e=OF;
model  LP1  / all /;
Solve  LP1  US LP  min  of ;
display  x1.l , x2.l , x3.l , of.l ;
```

The optimal solution is $\begin{bmatrix} x_1 \\ x_2 \\ x_3 \\ \text{OF} \end{bmatrix} = \begin{bmatrix} 0.333 \\ 1.333 \\ 3.667 \\ 15.333 \end{bmatrix}$. Clicking on the model statistic tap

shows that this model has four blocks of equations (four single equations). It has also four variables ($x_{1,2,3}$,OF). The solution report would be as follows:

S O L V E S U M M A R Y
MODEL LP1 OBJECTIVE of
TYPE LP DIRECTION MINIMIZE
SOLVER CPLEX FROM LINE 12
**** SOLVER STATUS 1 Normal Completion
**** MODEL STATUS 1 Optimal

(continued)

**** OBJECTIVE VALUE 15.3333
RESOURCE USAGE, LIMIT 0.016 1000.000
ITERATION COUNT, LIMIT 2 2000000000

It means that the solver has successfully solved the model and the solution is globally optimal. The solver used for solving the model is CPLEX [1]. It states that the value of objective function is 15.333. It also gives the user some info regarding the computational burden needed for solving the problem. RESOURCE USAGE is indicating how much time was needed to solve the model in seconds (0.016 s) and what was the maximum time allowed to do so (1000 s). The number of iterations needed for finding the optimal solution is two in this case. The default value for this limit is 2,000,000,000.

Clicking on the SolEQU tab would show the following info regarding the model:

			LOWER	LEVEL	UPPER	MARGINAL
—	EQU	eq1	3	3	+INF	0.333
—	EQU	eq2	5	5	+INF	2.333
—	EQU	eq3	4	4	4	0.667
—	EQU	eq4	0	0	0	-1

The lower limits of eq1,eq2 have some finite values (3,5) but their upper limits are $+\infty$. This means that these two equations are of $\geq$ type. Equations eq3,eq4 have equal values for lower and upper limits. This means that these equations are of equality type. The interesting part of the analysis is given in marginal column (the last column). As it can be seen, the level values of eq1,eq2,eq3 are equal to their lower limits. This has a certain meaning that these constraints are binding constraints. This means that if the lower limits are changed then the objective function value would change. The marginal values actually show the sensitivity coefficients of objective function to these equations. Let's check them in more detail. The marginal value of eq1 is 0.333. This means that $\frac{\Delta OF}{\Delta RHS\,eq1} = 0.333$. The right-hand side of eq1 is 3 so if it is increased to 3.2 then $\Delta RHS\,eq1 = 3.2 - 3 = 0.2$. The marginal value indicates that the new objective function would be $15.333 + 0.333 * 0.2 = 15.3996$. If the GAMS model is solved using the new RHS value of eq1 (3.2) then OF would be 15.4. The obtained value is close but not exactly what we were expecting but why? This is because the marginal values are accurate for very small change in RHS values of the equations. If the variations are small enough then the approximation would be accurate enough.

The marginal values constitute the values of dual variables. The decision maker can understand which constraint is binding (has nonzero marginal value) and also shows the most influential constraint on objective function (the biggest marginal value).

Clicking on the SolVAR tab would show the following info regarding the model:

			LOWER	LEVEL	UPPER	MARGINAL
——	VAR	x1	-INF	0.333	+INF	0
——	VAR	x2	-INF	1.333	+INF	0
——	VAR	x3	-INF	3.667	+INF	0
——	VAR	OF	-INF	15.333	+INF	0

The variable attributes are given in the table above. The lower and upper limits of all variables are $-\infty$ and $+\infty$, respectively. The marginal values are zero (this is because no bound is defined for variables). Suppose that we define a lower limit for x_2 which is $2 \leq x_2$. Let's see the impact on marginal values of equations and variables: The code would be as follows:

```
Variables x1,x2,x3,of;
Equations eq1,eq2,eq3,eq4;
eq1 .. x1+2*x2 =g=3;
eq2 .. x3+x2 =g=5;
eq3 .. x1+x3 =e=4;
eq4 .. x1+3*x2 +3*x3=e=OF;
Model LP1 /all/;
x2.lo=2;
Solve LP1 US LP min of;
display x1.l,x2.l,x3.l,of.l;
```

Clicking on the SolEQU tab would show the following info regarding the model:

			LOWER	LEVEL	UPPER	MARGINAL
——	EQU	eq1	3	5	+INF	0
——	EQU	eq2	5	5	+INF	2
——	EQU	eq3	4	4	4	1
——	EQU	eq4	0	0	0	-1

This table shows that eq1 is no longer the binding equation. This means that small change of RHS (which is 3 here) won't change the objective function. The marginal value of this equation is zero. The eq2 and eq3 are binding equations (they have nonzero marginal values and also their level is equal to their lower value).

Clicking on the SolVAR tab would show the following info regarding the model:

			LOWER	LEVEL	UPPER	MARGINAL
——	VAR	x1	-INF	1	+INF	0
——	VAR	x2	2	2	+INF	1
——	VAR	x3	-INF	3	+INF	0
——	VAR	OF	-INF	16	+INF	0

The marginal values of all variables are zero except $x2$ which is 1. This means that setting a lower limit for $x2$ caused this situation. Originally, the optimal value of $x2$ was 1.333 and now the lower limit (which is set to be 2) stops it from reaching its optimal value. Any possible decrease in $x2$ can help reducing the overall objective function.

GCode 2.2 Finding the boundaries of a variable example (2.3)

```
Variables  x1 , x2 , x3 , of ;
Equations
eq1 , eq2 , eq3 , eq4 ;
eq1  ..  x1+2*x2 =l =3;
eq2  ..  x3+x2 =l =2;
eq3  ..  x1+x2+x3 =e =4;
eq4  ..  x1+2*x2 −3*x3=e=OF;
Model LP1  / all / ;
x1 . lo =0;  x1 . up=5;  x2 . lo =0;  x2 . up=3;  x3 . lo =0;  x3 . up=2;
Solve LP1 US LP max of ;
display x1 . l , x2 . l , x3 . l , of . l ;
Solve LP1 US LP min of ;
display x1 . l , x2 . l , x3 . l , of . l ;
```

2.1.1.2 Boundary Determination Example

Sometimes it is needed to find the maximum and minimum of the objective function for a given model. This means that the problem should be solved two times. The example given in (2.3) is describing such a situation.

$$\min / \max_{X} \mathrm{OF} = x_1 + 2x_2 - 3x_3 \tag{2.3a}$$

$$x_1 + 2x_2 \leq 3 \tag{2.3b}$$

$$x_3 + x_2 \leq 2 \tag{2.3c}$$

$$x_1 + x_2 + x_3 = 4 \tag{2.3d}$$

$$0 \le x_1 \le 5 \tag{2.3e}$$

$$0 \le x_2 \le 3 \tag{2.3f}$$

$$0 \le x_3 \le 2 \tag{2.3g}$$

Please pay special attention to (2.3e)–(2.3g). These three constraints can be easily treated in GAMS using *.lo* and *.up* statements. In order to reduce the number of equations in the model it should be avoided defining them as six extra equations. The Gcode 2.2 for solving (2.3) is provided as follows:

The problem is solved and the solutions are obtained as $\begin{bmatrix} x_1 \\ x_2 \\ x_3 \\ OF \end{bmatrix} = \begin{bmatrix} 3 \\ 0 \\ 1 \\ 0 \end{bmatrix}_{max}$ and

$\begin{bmatrix} x_1 \\ x_2 \\ x_3 \\ OF \end{bmatrix} = \begin{bmatrix} 2 \\ 0 \\ 2 \\ -4 \end{bmatrix}_{min}$. The application of this model is in interval optimization [2], fuzzy optimization [3], and DC power flow (which will be discussed in Chap. 6)

2.1.2 Mixed Integer Programming (MIP)

In mixed integer programming (MIP) problems, the decision maker is faced with constraints and objective function that are linear but there exist some integer/binary variables.

2.1.2.1 MIP Example

A MIP example is given for clarification as follows:

GCode 2.3 MIP example (2.4)

```
Variables  x , of ;
Binary  variable  y ;
Equations eq1 ,  eq2 , eq3 ;
eq1  ..  −3*x+2*y  =g=1;
eq2  ..  −8*x+10*y  =l=10;
eq3  ..    x+y=e=OF;
Model MIP1  / all /;
x . up = 0.3;
Solve MIP1 US MIP max of ;
display y . l , x . l , of . l ;
```

$$\max_{x,y} \text{OF} = x + y \tag{2.4a}$$

$$- 3x + 2y \geq 1 \tag{2.4b}$$

$$- 8x + 10y \leq 10 \tag{2.4c}$$

$$y \in \{0, 1\}, 0.3 \leq x \tag{2.4d}$$

y is a binary variable and x is a real number. The GAMS code for solving (2.4) is provided in GCode 2.3:

By running the GAMS code the optimal solution is found as follows: $\begin{bmatrix} x \\ y \\ \text{OF} \end{bmatrix} =$

$\begin{bmatrix} 0.3 \\ 1 \\ 1.3 \end{bmatrix}_{\max}$

2.1.2.2 N-Queen Example

The N-queen problem is a classic MIP problem [4]. In this problem, it is tried to maximize the number of queens that can sit on a chessboard without attacking each other. The procedure is simple as follows: First of all, it is needed to define a variable x_{ij} which is a binary variable (0/1) and states whether the queen should sit (1) on block ij (row i, column j) or not (0). Additionally, if the queen is on block ij then no other queen can sit on row i or column j or the diagonal that contains cell ij. This is mathematically stated as follows:

$$\max_{x_{ij}} \text{OF} = \sum_{i,j} x_{ij} \tag{2.5a}$$

$$\sum_{i} x_{ij} \leq 1 \ \ \forall j \tag{2.5b}$$

$$\sum_{j} x_{ij} \leq 1 \ \ \forall i \tag{2.5c}$$

$$\sum_{c,r} x_{c,r} \leq 1 \ \ \forall i,j \in \left| \frac{i-r}{j-c} \right| = 1 \tag{2.5d}$$

If a queen is on a cell then no other queen can exist on the same column (2.5b) or the same row (2.5c) or the same diagonal (2.5d). The GAMS code for solving N-queen problem (2.5) is given in GCode 2.4:

The N-queen problem is solved two times in Gcode 2.4. Two models are defined in this code namely *MIP2a* and *MIP2b*. The Queen placement is solved on a 4×4 board, Fig. 2.1a shows the wrong placement. This solution is obtained from *MIP2a* and considers $eq_{1,2,3}$. This is because we are not considering diagonal movement of queen. In order to overcome this shortcoming two additional constraints $eq_{4,5}$ are considered in Model *MIP2b*. The correct solution is depicted in Fig. 2.1b.

It should be noted that the solution for the defined problem is optimal but not unique. This means that other configurations might be obtained that can satisfy the defined constraints. The developed GCode 2.4 is general and can be used for solving the problem for various sizes of the board. The optimal solution for queen placement on a (a) 8×8 chessboard and (b) 16×16 board is shown in Fig. 2.2.

GCode 2.4 N-queen example (2.5)

```
Sets i /1*4/, j /1*4/;
alias(i,row);
alias(j,col);
variable of;
binary variable x(i,j);
Equations eq1,eq2,eq3,eq4,eq5;
eq1(j) .. sum(i,x(i,j)) =l=1;
eq2(i) .. sum(j,x(i,j)) =l=1;
eq3    .. sum((i,j),x(i,j))=e=OF;
eq4(i,j) .. sum((row,col)$((ord(row)-ord(i))=(ord(col)-ord(j))),
     x(row,col))=l=1;
eq5(i,j) .. sum((row,col)$((ord(row)-ord(i))=-(ord(col)-ord(j)))
     ,x(row,col))=l=1;
Model MIP2a /eq1,eq2,eq3/;
Model MIP2b /all/;
Solve MIP2a US MIP max of;
display x.l,of.l;
Solve MIP2b US MIP max of;
display x.l,of.l;
```

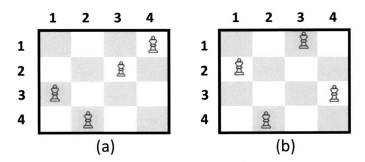

(a)   (b)

Fig. 2.1 Queen placement on a 4×4 board: (**a**) wrong placement, (**b**) optimal placement

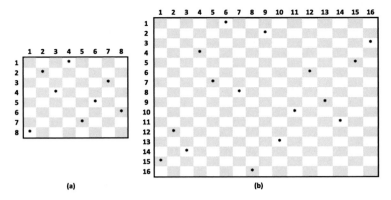

Fig. 2.2 Queen placement on a (**a**) 8 × 8 chessboard, (**b**) 16 × 16 board

2.1.2.3 Emergency Center Allocation

Consider six cities (1–6) which are located at different distances to each other. Each city should have access to an emergency center within a short period of time. The time required for moving from one city to another one is given in the following table in minutes.

	1	2	3	4	5	6
1	0	30	16	22	24	29
2		0	54	32	43	24
3			0	44	50	28
4				0	14	43
5					0	12
6						0

It should be noted that the values of this table are symmetrical. For example, distance from city 1 to city 2 is 30 min. It means that the distance from city 2 to city 1 is also 30 min. The critical time for reaching to the emergency center is assumed to be 20 min. The question is what is the minimum number of cities that should host emergency center? Which cities should be chosen?

By observing the first row of distance matrix, it is understood that if city 1 or city 6 have the emergency center then city 1 meets the access requirement. The same concept applies for city 2. City 2 should definitely have an emergency center because no other city is located within 20 min distance of this city. If the allocation decision is defined as a binary variable x_i then the following inequality should be satisfied for city 1:

$$x_1 + x_6 \geq 1 \qquad (2.6)$$

The following inequality should be satisfied for city 2:

$$x_2 \geq 1 \tag{2.7}$$

The overall constraints are as follows:

$$x_1 + x_6 \geq 1 \tag{2.8}$$

$$x_2 \geq 1 \tag{2.9}$$

$$x_3 + x_5 \geq 1 \tag{2.10}$$

$$x_4 + x_5 \geq 1 \tag{2.11}$$

$$x_3 + x_4 + x_5 + x_6 \geq 1 \tag{2.12}$$

$$x_1 + x_5 + x_6 \geq 1 \tag{2.13}$$

Two different approaches will be presented here for modeling this problem.
Scalar equations:

In this approach, we initially analyzed the distance data and understood what kind
of relations should be enforced for different variables. The GCode 2.5 is describing
how to do this.

GCode 2.5 Scalar equations for emergency centre allocation

```
binary  variable  x1 , x2 , x3 , x4 , x5 , x6 ;
variable  OF;
equations
eq1 , eq2 , eq3 , eq4 , eq5 , eq6 , eq7 ;
eq1  ..  x1+x6  =g=1;
eq2  ..  x2  =g=1;
eq3  ..  x3+x5  =g=1;
eq4  ..  x4+x5  =g=1;
eq5  ..  x3+x4+x5+x6  =g=1;
eq6  ..  x1+x5+x6  =g=1;
eq7  ..  x1+x2+x3+x4+x5+x6  =e=OF;
Model  emergency  / all / ;
Solve  emergency  us  mip  min  of ;
```

The optimal answer is OF = 3 (three cities should host emergency centers). The
candidate cities are x_1, x_2, x_5. The problem with this kind of modeling is that it needs
pre-processing of the raw data and also in case, the number of cities are changed then
it is need to extensively modify the code. A much more efficient way of coding this
problem is using the extended equations.
Indexed equations:

GCode 2.6 Indexed equations for emergency centre allocation

```
set  city  /1*6/          ;
alias(city,town);
binary  variable  x(city);
variable  OF;
table  data(city,town)
        1     2     3     4     5     6
1       0     30    46    22    24    19
2             0     54    32    43    24
3                   0     44    16    28
4                         0     14    43
5                               0     12
6                                     0;
data(city,town)$data(town,city)=data(town,city);
scalar  criticaltime  /20/;
Equations
eq1,eq2;
eq1(city)  ..  sum(town$(data(city,town)<criticaltime),  x(town))  =g
    =1;
eq2  ..  OF=e=sum(city,x(city));
Model  emergency  /all/;
Solve  emergency  us  mip  min  of;
```

The developed code will provide the same answer as before but it has the following features:

- It is not needed to manually write one equation for each constraint.
- It works for any number of cities.
- The distance data is fed to the model using a table. This will be useful for cases that the input data might change.
- Debugging and tracing the code are much easier for the users.
- The code does not change if the critical access time is updated.

The following line of the code is to make the data matrix symmetrical.

```
data(city,town)$data(town,city)=data(town,city);
```

The following line describes the condition for accessing the emergency center (for each city). The equation eq1 is defined over the set "city."

```
eq1(city) .. sum(town$(data(city,town)<criticaltime), x(town)) =g=1;
```

The optimal solution is:

```
—— VAR x
LOWER LEVEL UPPER MARGINAL
1 . . 1.000 1.000
2 . 1.000 1.000 1.000
3 . . 1.000 1.000
4 . . 1.000 1.000
5 . 1.000 1.000 1.000
6 . 1.000 1.000 1.000
```

2.1.3 Nonlinear Programming (NLP)

In nonlinear programming problems, at least one of f, G, H in (2.1) is nonlinear.

2.1.3.1 NLP Example (2.14)

$$\max_{x_i} \text{OF} = x_1 x_4 (x_1 + x_2 + x_3) + x_2 \tag{2.14a}$$

$$x_1 x_2 x_3 x_4 \geq 20 \tag{2.14b}$$

$$x_1^2 + x_2^2 + x_3^2 + x_4^2 = 30 \tag{2.14c}$$

$$1 \leq x_1, x_2, x_3, x_4 \leq 3 \tag{2.14d}$$

The GAMS code for solving example (2.14) is given in GCode 2.7:

GCode 2.7 Example (2.14)

```
variable of , x1 , x2 , x3 , x4 ;
equations
eq1 , eq2 , eq3 ;
eq1  ..  x1*x4*(x1+x2+x3)+x2=e=OF;
eq2  ..  x1*x2*x3*x4 =g=20;
eq3  ..  x1*x1+x2*x2+x3*x3+x4*x4=e=30;
x1.lo=1;
x1.up=3;
x2.lo=1;
x2.up=3;
x3.lo=1;
x3.up=3;
x4.lo=1;
x4.up=3;
Model NLP1 / all /;
Solve NLP1 US NLP max of ;
```

The solution for example (2.14) obtained by GCode 2.7 is as follows:

Variables	Lower	Level	Upper	Marginal
of	$-\infty$	73.605	$+\infty$	0
$x1$	1.000	3.000	3.000	21.025
$x2$	1.000	2.575	3.000	0
$x3$	1.000	2.317	3.000	0
$x4$	1.000	3.000	3.000	12.025

It is worth noting that providing a starting point for the variables can help the GAMS in finding a better solution. Generally speaking, finding the optimal solution of the model is not guaranteed in NLP problems. The initial values for variables are set by "X.l=initial value" command before solve statement.

2.1.3.2 Circle Placement Example (2.15)

Suppose there are n circles with known radius values (R_i). The question is: what is the minimum surface of the table that these circles can be placed on it without any overlapping. The concept of optimal circle placement on a given table is shown in Fig. 2.3. The decision variables of this problem are table dimensions $\{w, h\}$ and center locations $\{x_i, y_i\}$. The objective function is defined as the surface of the table (OF $= w \times h$). The constraints are described as:

- Each circle should be completely on the table. This requires: $R_i \leq x_i \leq w - R_i$ and similarly $R_i \leq y_i \leq h - R_i$.
- For every two circles, no overlap should happen. This means that: $(x_i - x_j)^2 + (y_i - y_j)^2 \geq (R_i + R_j)^2$

The circle placement problem is formulated as:

$$\min_{x_i, y_i, w, h} \text{OF} = hw \tag{2.15a}$$

$$R_i \leq x_i \leq w - R_i \tag{2.15b}$$

$$R_i \leq y_i \leq h - R_i \tag{2.15c}$$

$$(x_i - x_j)^2 + (y_i - y_j)^2 \geq (R_i + R_j)^2 \tag{2.15d}$$

It is assumed that R_i are known in advance and they are treated as input data. It is also assumed that there are six circles available. The GAMS code for solving circle placement example (2.15) is as GCode 2.8:

GCode 2.8 Circle placement example (2.15)

```
set i /1*6/; alias(i,j);
Positive variables x(i),y(i),w,h; variable of;
parameter radius(i)
/1  2
2  1.2
3  1.8
4  0.9
5  3.2
6  0.7/;
Equations eq1,eq2,eq3,eq4;
eq1 .. w*h=e=OF;
eq2(i) .. x(i)=l=w-radius(i);
eq3(i) .. y(i)=l=h-radius(i);
eq4(i,j)$(ord(i)<>ord(j))..power(y(j)-y(i),2)+power(x(j)-x(i),2)=
    g=(radius(i)+radius(j),2);
x.lo(i)=radius(i); y.lo(i)=radius(i);
Model NLP1 /all/;
Solve NLP1 US NLP min of;
parameter report(i,*);
report(i,'X')=x.l(i); report(i,'y')=y.l(i);
report(i,'R')=radius(i);
display report,of.l,w.l,h.l;
```

As it is observable in GCode 2.8, the radius of circles are given as parameters. The constraint given in (2.15b) is actually two constraints $R_i \leq x_i$ and $x_i \leq w - R_i$. The first one should be treated using *.lo* statement and the second one should be modeled using equations. This is because it involves two variables x_i, w which should be valid for every member of set i. This is why the equation $eq2$ is defined over set i.

The optimal solution of circle placement on a table is given in Fig. 2.4.

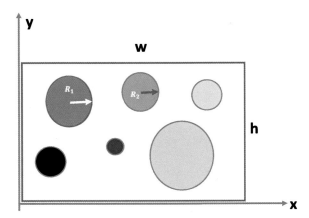

Fig. 2.3 Optimal circle placement on a given surface

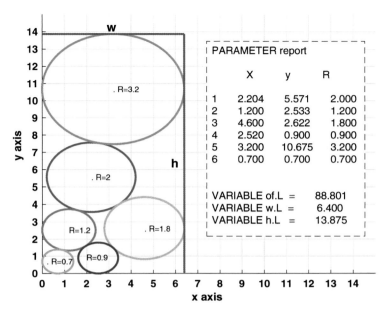

Fig. 2.4 The optimal solution of circle placement on a surface

2.1.3.3 Maximizing the Area of a Right Triangle with Constant Circumference

Suppose that we have some limited meters of wire fences (*C*) and are requested to enclose an area (right triangle shape). Determine the dimensions of such a triangle. Considering Fig. 2.5, the following optimization should be solved:

$$\max_{H,B} \text{OF} = \frac{H * B}{2} \tag{2.16a}$$

$$H + B + \sqrt{H^2 + B^2} = C \tag{2.16b}$$

where *C* is the length of the available wire fence. *H* and *B* are the height and base of the triangle, respectively. The GCode 2.9 provides the solution for maximizing the area of a triangle with constant circumference.

GCode 2.9 Maximizing the area of a triangle

```
Positive  variables  h, b;
Variable  OF;
Scalar  C  /15/;  h.up=C;  b.up=C;
Equations  eq1,eq2;
eq1  ..  h+b+sqrt(h*h+b*b)=e=C;
eq2  ..  OF=e=0.5*h*b;
Model  triangle  /all/;
Solve  triangle  us  nlp  min  of;
```

Fig. 2.5 Maximizing the
area of a triangle with
constant circumference

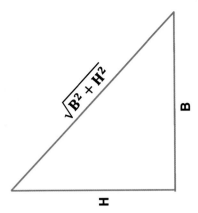

2.1.4   Quadratic Constrained Programming (QCP)

The quadratic programming is a special case of NLP problems. In QCP problems, at least one of f, G, H in (2.1) is nonlinear but nonlinearity is of quadratic form.

2.1.4.1   QCP Example (2.17)

Consider the following QCP optimization problem:

$$\max_{x_i} \text{OF} = -3x_1^2 - 10x_1 + x_2^2 - 3x_2 \tag{2.17a}$$

$$x_1 + x_2^2 \geq 2.5 \tag{2.17b}$$

$$2x_1 + x_2 = 1 \tag{2.17c}$$

The GAMS code for solving example (2.17) is as GCode 2.10:

GCode 2.10   QCP Example (2.17)

```
variable of, x1, x2;
equations
eq1, eq2, eq3;
eq1  ..  −3*x1*x1  −10*x1  +x2*x2−3*x2=e=OF;
eq2  ..  x1+x2*x2  =g=2.5;
eq3  ..  2*x1+x2=e=1;
Model QCP1 / all /;
Solve QCP1 US QCP max of;
```

The solution for example (2.17) obtained by GCode 2.10 is as follows:

Variables	Lower	Level	Upper	Marginal
of	$-\infty$	-9.550	$+\infty$	0
$x1$	$-\infty$	1.093	$+\infty$	0
$x2$	$-\infty$	-1.186	$+\infty$	0

2.1.4.2 QCP Example (2.18)

Another simple QCP problem is described as follows:

$$\min_{x_i} OF = x_1^2 - 10x_1 + x_2^2 - 3x_2 \tag{2.18a}$$

$$x_1^2 + x_2 \leq 5 \tag{2.18b}$$

$$2x_1 - x_2 \geq 1 \tag{2.18c}$$

GCode 2.11 Example (2.18)

```
Variable of ,x1 ,x2 ;
Equations eq1 ,eq2 ,eq3 ;
eq1  ..  x1*x1 −10*x1 +x2*x2−3*x2=e=OF;
eq2  ..  x1*x1+x2 =l=5;
eq3  ..  2*x1−x2=g=1;
Model QCP1 / all /;
Solve QCP1 US QCP min of ;
```

The solution for example (2.18) (obtained by GCode 2.11) is as follows:

Variables	Lower	Level	Upper	Marginal
of	$-\infty$	-18.054	$+\infty$	0
$x1$	$-\infty$	2.054	$+\infty$	0
$x2$	$-\infty$	0.783	$+\infty$	0

2.1.5 Mixed Integer Nonlinear Programming (MINLP)

In MINLP problems, at least one of f, G, H in (2.1) is nonlinear and integer variables are involved.

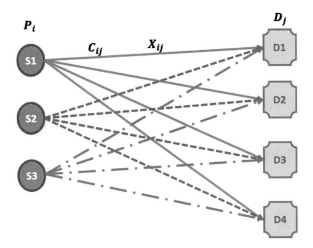

Fig. 2.6 Optimal transportation problem

2.1.5.1   Minimum Transportation Cost Example (2.19)

The transportation problem is a classic example which has been modified for this example. There are some suppliers (node i) and some demands (node j) which should be supplied. The problem is to determine how much each supplier should produce and how to transfer them to the demand points. The transportation costs should be minimized.

The cost of transportation is assumed to be related to the square of quantity transported from node i to node j. The transportation costs are proportional to the square of product transported from i to j and the length of the rout. The object is minimizing the total transportation costs and supplying all demand in different nodes. The optimal transportation problem is depicted in Fig. 2.6. The cost coefficient and maximum flow of each rout (C_{ij}), demand (D_j), and capacity of each producer are known.

GCode 2.12   Example (2.19)

```
sets
i  /s1*s3/
j  /D1*D4/;
table  C(i,j)
      d1              d2              d3              d4
s1  0.0755          0.0655          0.0498          0.0585
s2  0.0276          0.0163          0.096           0.0224
s3  0.068           0.0119          0.034           0.0751;
table  data(i,*)
      'Pmin'    'Pmax'
s1     100      450
```

```
s2    50        350
s3    30        500;
parameter  demand(j)
/d1  217
d2  150
d3  145
d4  244/;
variable  of,x(i,j),P(i);
binary  variable  U(i);
equations
eq1,eq2,eq3,eq4,eq5;
eq1  ..  OF=e=sum((i,j),C(i,j)*x(i,j)*x(i,j));
eq2(i)  ..     P(i)=l=data(i,'Pmax')*U(i);
eq3(i)  ..     P(i)=g=data(i,'Pmin')*U(i);
eq4(j)  ..  sum(i,x(i,j))=g=demand(j);
eq5(i)  ..  sum(j,x(i,j))=e=P(i);
P.lo(i)=0;
P.up(i)=data(i,'Pmax');
x.lo(i,j)=0;
x.up(i,j)=100;
Model  minlp1  /all/;
Solve  minlp1  US  minlp  min  of;
```

The optimal transportation problem is formulated in (2.19).

$$\min_{x_{ij}} \text{OF} = \sum_{ij} C_{ij} x_{ij}^2 \tag{2.19a}$$

$$\begin{cases} P_i^{\min} \leq P_i \leq P_i^{\max}, & \text{if unit } i \text{ is on,} \\ P_i = 0 & \text{if unit } i \text{ is off} \end{cases} \tag{2.19b}$$

$$\sum_i x_{ij} \geq D_j \tag{2.19c}$$

$$\sum_j x_{ij} = P_i \tag{2.19d}$$

$$0 \leq x_{ij} \leq x_{ij}^{\max} \tag{2.19e}$$

The road flow limit is assumed to be $x_{ij}^{\max} = 100$.

i	$P_i^{\min}$	$P_i^{\max}$	j			
	100	450	0.0755	0.0655	0.0498	0.0585
	50	350	0.0276	0.0163	0.096	0.0224
	30	500	0.068	0.0119	0.034	0.0751
D_j			217	150	145	244

The GAMS code for solving example (2.19) is as GCode 2.12:
The solution for example (2.19) obtained by GCode 2.12 is as follows:

P_i	X_{ij}			
199.245	55.443	14.255	48.601	80.946
282.494	100	57.282	25.212	100
274.261	61.557	78.463	71.187	63.054

2.1.5.2 Benefit Maximization Transportation Example (2.20)

Reconsider the optimal transportation problem formulated in (2.19). This problem
is a minimum cost problem and the goal is service provision to the consumers. Now
suppose that the objective is benefit maximization. The machine (i) will produce
equal to P_i and sell it to different consumer j. The maximum value of purchase that
a consumer may procure is D_j. The revenue that is obtained from selling product is
k \$/unit. Now the benefit maximization is modeled as follows:

$$\max_{x_{ij}} OF = \sum_i P_i k - \sum_{ij} C_{ij} x_{ij}^2 \tag{2.20a}$$

$$\begin{cases} P_i^{\min} \le P_i \le P_i^{\max}, & \text{if unit } i \text{ is on}, \\ P_i = 0 & \text{if unit } i \text{ is off} \end{cases} \tag{2.20b}$$

$$\sum_i x_{ij} \le D_j \tag{2.20c}$$

$$\sum_j x_{ij} = P_i \tag{2.20d}$$

$$0 \le x_{ij} \le x_{ij}^{\max} \tag{2.20e}$$

The GAMS code for solving the example (2.20) is given in GCode 2.13.
The solution for example (2.20) obtained by GCode 2.13 is as follows:

	P_i	X_{ij}			
i	0	0	0	0	0
	137.377	32.609	55.215	9.375	40.179
	0	0	0	0	0

2.2   Random Numbers in GAMS

Random number generation in GAMS is a simple task. There are different built-in probability density function that can be used for generating random numbers. Some popular random number generators are listed as follows:

- Beta function

$$\frac{\Gamma(\alpha + \beta)}{\Gamma(\alpha)\Gamma(\beta)}(1 - x)^{\beta-1}(x)^{\alpha-1} \tag{2.21}$$

The format of this function in GAMS is beta(α,β)

- Uniform function (continuous)

$$P(x) = \begin{cases} \frac{1}{(b-a)}, & \text{for } a \leq x \leq b \\ 0, & \text{otherwise} \end{cases} \tag{2.22}$$

The format of this function in GAMS is uniform(a, b)
- Uniform function (discrete)

$$P(x) = \begin{cases} \frac{1}{N}, & \text{for } a \leq x \leq b \\ 0, & \text{otherwise} \end{cases} \tag{2.23}$$

GCode 2.13 Example (2.20)

```
sets
i  /s1*s3/
j  /D1*D4/;
scalar k /1.8/;
table C(i,j)
      d1              d2              d3            d4
s1  0.0755          0.0655          0.0498        0.0585
s2  0.0276          0.0163          0.096         0.0224
s3  0.068           0.0119          0.034         0.0751;
table  data(i,*)
     'Pmin'   'Pmax'
s1    100      450
s2    50       350
s3    30       500;
parameter demand(j)
/d1  217
d2  150
d3  145
d4  244/;
variable  of,x(i,j),P(i);
binary  variable  U(i);
```

```
equations
eq1 , eq2 , eq3 , eq4 , eq5 ;
eq1   ..   OF=e=sum(i,k*P(i))−sum((i,j),C(i,j)*x(i,j)*x(i,j));
eq2(i)   ..     P(i)=l=data(i,'Pmax')*U(i);
eq3(i)   ..     P(i)=g=data(i,'Pmin')*U(i);
eq4(j)   ..   sum(i,x(i,j))=l=demand(j);
eq5(i)   ..   sum(j,x(i,j))=e=P(i);
P.lo(i)=0;
P.up(i)=data(i,'Pmax');
x.lo(i,j)=0;
x.up(i,j)=100;
Model minlp2 /all/;
Solve minlp2 US minlp max of;
```

where N is the total integer numbers in $[a, b]$ interval. The format of this function in GAMS is uniformint(a, b)

2.2.1 Estimating the π Number

Calculating the π number is investigated and coded in GAMS. Consider a circle inscribed in a square as shown in Fig. 2.7. If a point is randomly dropped on this square then the probability of sitting on the circle would be $\frac{\text{Circle area}}{\text{Square area}}$. In order to calculate the π number a simple experience is done. The point is dropped on the square area for N times. The number of events that the point is on circle area would be n. The following equation would be valid if N is a big number. The GAMS code for solving this problem is described in GCode 2.14.

$$\frac{\text{Circle area}}{\text{Square area}} = \frac{n}{N} \qquad (2.24)$$

$$\frac{\pi R^2}{4R^2} = \frac{n}{N} \qquad (2.25)$$

GCode 2.14 π number estimation

```
scalar low /0/, High /1/, pistimate;
set counter /c1*c200/;
parameter report(counter,*);
report(counter,'x')=uniform(LOW,HIGH);
report(counter,'y')=uniform(LOW,HIGH);
pistimate=4*sum(counter$(power(report(counter,'x')
−0.5,2)+power(report(counter,'y')−0.5,2)<=0.25),1)/card(counter);
display report , pistimate;
```

As it can be seen in GCode 2.14, there is no solve statement needed. This is because no variable is defined in the model and no optimization is going to take place.

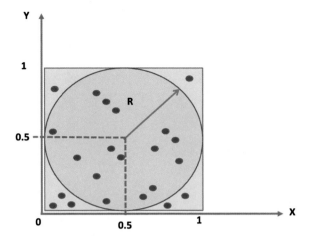

Fig. 2.7 Estimating the π number

2.2.2 Integration Calculation

The random numbers can be used for calculating the integration problems. A simple example is given as follows:

$$Z = \int_0^1 1 - x\sin(20x)dx \tag{2.26}$$

In order to calculate the area under the graph as it is shown in Fig. 2.8, the following steps are followed:

- Set Counter = 1; $n = 1$;
- Generate a pair of random numbers (x, y) where $0 \leq X \leq 1$ and $0 \leq Y \leq 2$.
- Check if $y \leq f(X)$
- Set Counter = Counter + 1;

This procedure is repeated for max number of counter (N). For large values of N, the Z can be calculated using the following relation:

$$\frac{Z}{2*1} = \frac{n}{N} \tag{2.27}$$

The following command should be added to the code (before calling the uniform function) in order to have a set of new random numbers every time the code is run:

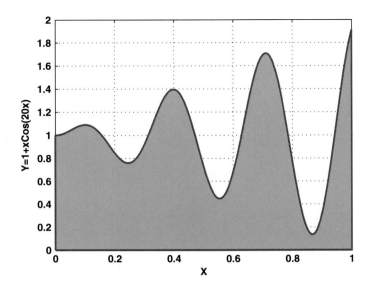

Fig. 2.8 Numerical integration using random numbers

GCode 2.15 Integration calculation using random numbers

```
Scalar  Zstimate ;
Set  counter  /c1*c200000/;
parameter  report(counter ,*);
report(counter , 'x')=uniform(0,1);
report(counter , 'y')=uniform(0,2);
Zstimate=2*sum(counter$(report(counter , 'y')<1+
report(counter , 'x')*sin(report(counter , 'x')*20)),1)/card(counter);
Sisplay  report  ,  Zstimate ;
```

    execseed = 1+gmillisec(jnow);

Further info on how to use random numbers in GAMS can be found in [5].

2.2.3 LP Problems with Uncertain Coefficients

Consider the following LP problem:

$$Z = \max_{x_1, x_2} 750x_1 + 1000x_2 \tag{2.28}$$

$$x_1 + x_2 \leq a \tag{2.29}$$

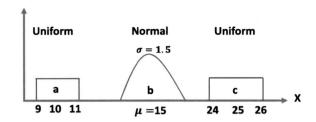

Fig. 2.9 The probability density function for each uncertain parameter

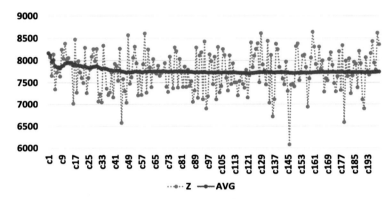

Fig. 2.10 Uncertain coefficients in LP problems

$$x_1 + 2x_2 \leq b \tag{2.30}$$

$$4x_1 + 3x_2 \leq c \tag{2.31}$$

$$x_1 \geq 0 \tag{2.32}$$

$$x_2 \geq 0 \tag{2.33}$$

where a, b, and c are uncertain random parameters. The probability density function for each parameter is given in Fig. 2.9. The question is how to describe the probability density of Z? If all uncertain parameters (a,b,c) are equal to their expected values $(10,15,25)$ then the objective function would be 7750. The following steps are followed to find out the distribution of Z in case of uncertain (a,b,c):

- Set Counter = 1;
- Generate a sample of random parameters (a,b,c) using the specified probability distribution functions
- Calculate the optimal Z and save the (a,b,c,x_1,x_2,Z).

The variation of Z along with the average value of Z are plotted in Fig. 2.10. The graph shows how the average value of Z converges.

2.3 Multi-Objective Optimization

In Multi-Objective Decision Making (MODM) methods, the decision alternatives are found considering the constraints of the given problem. In most practical optimization problems, particularly those applicable in power system, there exists more than one objective function which should be optimized simultaneously. These objectives functions might be in conflict, interdependent or independent of each other, so it is impossible to satisfy them all at once. The main differences between the multi-objective optimization and traditional single optimization techniques can be categorized into two groups:

1. Several objective functions are to be optimized at the same time.
2. There exists a set of optimal solutions which are mathematically equally good solutions (it means any of them cannot be preferred against others and a trade-off should be made to select one) instead of a unique optimal solution

2.3.1 Weighted Sum Approach

Some methods try to convert the multi-objective optimization problems into single objective one. However, this approach is not always applicable especially in the following cases:

• The objectives are not of the same type. For example voltage deviation and cost, weight and volume, surface and time.
• The weight coefficients in weighted sum approach cannot be easily determined. Additionally, If decision maker's preference is changed then the problem should be solved again.
• The single objective approach provides just one solution to the decision maker. It cannot show the tradeoff between two objective functions.
• The objective function is of the same type but there are multiple decision makers with different preferences. Each decision maker is trying to optimize its own objective function.

2.3.2 Pareto Optimality

The notion of optimality has been redefined in this context and instead of aiming to find a single solution, it is tried to produce a set of acceptable compromises or trade-offs from which the decision maker can choose one. The set of all optimal solutions which are non-dominated by any other solution is known as Pareto-optimal set. Suppose a minimizing problem with two objectives in conflict the Pareto optimal fronts are plotted in Fig. 2.11.

For every two solution in each Pareto front (like A, B) none of them is better than the others considering all objective functions. Here, f_1 is better minimized in solution A compared to B and f_2 is better minimized in solution B compared to A.

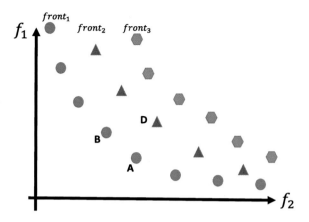

Fig. 2.11 Classification of a population to k non-dominated fronts

The same concept also applies for solutions in other fronts. For every solution in Pareto front k (for example D in the second front) there exists at least one solution in front $k-1$ (here A in the first front) that dominates it (is better considering all objective functions). Since solutions A and B belong to the first front, there is no solution better than them in respect to all objectives. Consider the bi-objective optimization described in (2.34):

$$\min_x f_1, f_2 \tag{2.34}$$

$$\text{Constraints}$$

Each solution in Pareto optimal set has two basic characteristics:

1. For every two solutions belonging to the same Pareto front (2.35) holds:

$$\forall i \exists j, n | f_n(\bar{x}_i) > f_n(\bar{x}_j) \tag{2.35}$$

$$\bar{x}_j, \bar{x}_i \in S$$

This means that for every solution X_i belonging to Pareto front S, at least one solution exists as X_j which is better than X_i at least in one objective function (named n here). In other words, there is no solution in Pareto optimal front which is the best among all members of this set considering all objectives.

2. For every solution belonging to an upper Pareto front and the ones in the lower fronts, (2.36) holds:

$$\forall k \in \{1 \ldots N_O\} f_k(\bar{x}_1) \le f_k(\bar{x}_2) \tag{2.36}$$

$$\exists k' \in \{1 \ldots N_O\} f_{k'}(\bar{x}_1) < f_{k'}(\bar{x}_2) \tag{2.37}$$

$$\bar{x}_1 \in S, \bar{x}_2 \in S^*$$

$$S < S^*$$

This means for every solution belonging to an upper Pareto front, there exists at least one solution in a lower Pareto front which is not worse in any objective function and is better in at least one objective function. N_O is the number of objective functions.

The classic approach for finding the Pareto optimal set is preference-based method in which a relative preference vector is used to weight the objectives and change them into a scalar value. By converting a multi-objective optimization problem into a single objective one, only one optimum solution can be achieved which is very sensitive to the given weights. The GAMS structure can solve only one objective function at once. This means it is needed to solve the multi-objective problem several times to obtain the Pareto optimal front. For this purpose, consider the bi-objective problem described in (2.38):

$$\max_{x_{1,2}} f_1 = 4x_1 - 0.5x_2^2 \tag{2.38a}$$

$$\max_{x_{1,2}} f_2 = -x_1^2 + 5x_2 \tag{2.38b}$$

$$2x_1 + 3x_2 \leq 10 \tag{2.38c}$$

$$2x_1 - x_2 \geq 0 \tag{2.38d}$$

$$1 \leq x_1 \leq 2 \tag{2.38e}$$

$$1 \leq x_1 \leq 3 \tag{2.38f}$$

GCode 2.16 Pareto optimal front example (2.38)

```
set counter /c1*c21/;
scalar E;
parameter report(counter,*);
variable of1,of2,x1,x2;
equations
eq1,eq2,eq3,eq4;
eq1 .. of1=e=4*x1-0.5*x2*x2;
eq2 .. of2=e=-x1*x1+5*x2;
eq3 .. 2*x1+3*x2=l=10;
eq4 .. 2*x1-x2=g=0;
x1.lo=1;
x1.up=2;
x2.lo=1;
x2.up=3;
Model pareto1 /all/;
parameter ranges(*);
Solve pareto1 US nLP max of1;
ranges('OF1max')=of1.l;
ranges('OF2min')=of2.l;
Solve pareto1 US nLP max of2;
ranges('OF2max')=of2.l;
ranges('OF1min')=of1.l;
```

```
E=ranges('OF1min');
loop(counter,
E=(ranges('OF2max')−ranges('OF2min'))*(ord(counter)−1)/(card
    (counter)−1)+ranges('OF2min');
of2.lo=E;
Solve pareto1 US nLP max of1;
report(counter,'OF1')=OF1.1;
report(counter,'OF2')=OF2.1;
report(counter,'E')=E;
);
Display report;
```

In this problem, two objectives should be maximized simultaneously. The procedure is as follows:

1. Find the maximum of each objective function and save them.
2. Add one of the objective functions to the constraints as follows:

$$f_2 \geq \epsilon \qquad (2.39)$$

The ϵ value will be varied from f_2^{min} to f_2^{max} and the f_1 is maximized.
The solution for example (2.38) obtained by GCode 2.16 is shown in Fig. 2.12.

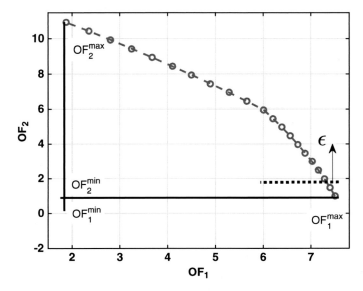

Fig. 2.12 The Pareto optimal front for bi-objective problem

2.4 Applications

Some optimization models used in power system studies are given in this section:

- LP programming: Transmission network estimation [6], short-term hydro scheduling [7], relay coordination [8], security constrained economic dispatch [9].
- MIP: Optimal PMU placement [10], unit commitment [11], Phase shifter placement in large-scale systems [12], minimum-losses radial configuration of electrical distribution networks [13].
- QCP: Topology identification in distribution network [14], optimum active and reactive generation dispatch [15].
- NLP: Voltage stability security margin calculation [16], reactive power planning [17], OPF [18].
- MINLP: Unit commitment with AC OPF constraints [19], flexible transmission expansion planning with uncertainties in an electricity market [20], optimal DG allocation [21].
- Multi-objective optimization: Transmission expansion planning [22], generation expansion planning [23], distribution network planning [24].

References

1. A. Meeraus, A. Brooke, D. Kendrick, R. Raman, *GAMS/Cplex 7.0 User Notes* (GAMS Development Corporation, Washington, DC, 2000)
2. Y. Wang, Q. Xia, C. Kang, Unit commitment with volatile node injections by using interval optimization. IEEE Trans. Power Syst. **26**(3), 1705–1713 (2011)
3. A. Soroudi, Possibilistic-scenario model for DG impact assessment on distribution networks in an uncertain environment. IEEE Trans. Power Syst. **27**(3), 1283–1293 (2012)
4. X. Hu, R.C. Eberhart, Y. Shi, Swarm intelligence for permutation optimization: a case study of n-queens problem, in *Swarm Intelligence Symposium, 2003. SIS'03. Proceedings of the 2003 IEEE* (IEEE, Piscataway, 2003), pp. 243–246
5. E. Kalvelagen, Some notes on random number generation with GAMS, 2005
6. L.L. Garver, Transmission network estimation using linear programming. IEEE Trans. Power Apparatus Syst. **PAS-89**(7), 1688–1697 (1970)
7. G.W. Chang, M. Aganagic, J.G. Waight, J. Medina, T. Burton, S. Reeves, M. Christoforidis, Experiences with mixed integer linear programming based approaches on short-term hydro scheduling. IEEE Trans. Power Syst. **16**(4), 743–749 (2001)
8. B. Chattopadhyay, M.S. Sachdev, T.S. Sidhu, An on-line relay coordination algorithm for adaptive protection using linear programming technique. IEEE Trans. Power Delivery **11**(1), 165–173 (1996)
9. R.A. Jabr, A.H. Coonick, B.J. Cory, A homogeneous linear programming algorithm for the security constrained economic dispatch problem. IEEE Trans. Power Syst. **15**(3), 930–936 (2000)
10. B. Gou, Generalized integer linear programming formulation for optimal PMU placement. IEEE Trans. Power Syst. **23**(3), 1099–1104 (2008)
11. J. Ostrowski, M.F. Anjos, A. Vannelli, Tight mixed integer linear programming formulations for the unit commitment problem. IEEE Trans. Power Syst. **27**(1), 39–46 (2012)

12. F.G.M. Lima, F.D. Galiana, I. Kockar, J. Munoz, Phase shifter placement in large-scale systems via mixed integer linear programming. IEEE Trans. Power Syst. **18**(3), 1029–1034 (2003)

13. A. Borghetti, A mixed-integer linear programming approach for the computation of the minimum-losses radial configuration of electrical distribution networks. IEEE Trans. Power Syst. **27**(3), 1264–1273 (2012)

14. Z. Tian, W. Wu, B. Zhang, A mixed integer quadratic programming model for topology identification in distribution network. IEEE Trans. Power Syst. **31**(1), 823–824 (2016)

15. H. Nicholson, M.J.H. Sterling, Optimum dispatch of active and reactive generation by quadratic programming. IEEE Trans. Power Apparatus Syst. **PAS-92**(2), 644–654 (1973)

16. L.A.L. Zarate, C.A. Castro, J.L.M. Ramos, E.R. Ramos, Fast computation of voltage stability security margins using nonlinear programming techniques. IEEE Trans. Power Syst. **21**(1), 19–27 (2006)

17. L.L. Lai, J.T. Ma, Application of evolutionary programming to reactive power planning-comparison with nonlinear programming approach. IEEE Trans. Power Syst. **12**(1), 198–206 (1997)

18. J.A. Momoh, R. Adapa, M.E. El-Hawary, A review of selected optimal power flow literature to 1993. I. nonlinear and quadratic programming approaches. IEEE Trans. Power Syst. **14**(1), 96–104 (1999)

19. A. Castillo, C. Laird, C.A. Silva-Monroy, J.P. Watson, R.P. ONeill, The unit commitment problem with ac optimal power flow constraints. IEEE Trans. Power Syst. **31**(6), 4853–4866 (2016)

20. J.H. Zhao, Z.Y. Dong, P. Lindsay, K.P. Wong, Flexible transmission expansion planning with uncertainties in an electricity market. IEEE Trans. Power Syst. **24**(1), 479–488 (2009)

21. A. Kumar, W. Gao, Optimal distributed generation location using mixed integer non-linear programming in hybrid electricity markets. IET Gener. Transm. Distrib. **4**(2), 281–298 (2010)

22. P. Maghouli, S.H. Hosseini, M.O. Buygi, M. Shahidehpour, A scenario-based multi-objective model for multi-stage transmission expansion planning. IEEE Trans. Power Syst. **26**(1), 470–478 (2011)

23. S. Kannan, S. Baskar, J.D. McCalley, P. Murugan, Application of NSGA-II algorithm to generation expansion planning. IEEE Trans. Power Syst. **24**(1), 454–461 (2009)

24. V. Miranda, J.V. Ranito, L.M. Proenca. Genetic algorithms in optimal multistage distribution network planning. IEEE Trans. Power Syst. **9**(4), 1927–1933 (1994)

Chapter 3
Power Plant Dispatching

This chapter provides the instruction on how to model the economic dispatch problem of different power plants. Different power plant technologies will be discussed, such as thermal power, wind turbine, CHP, and hydro power plants in GAMS.

The idea of dispatching generating units is based on some assumptions as follows:

- The technical characteristics of the units are known.
- The on/off status of each unit is known.
- It is solved for a single snapshot of the demand.

3.1 Thermal Unit Economic Dispatch

The thermal unit technology transforms the fuel-based source of energy into electricity. The production costs of thermal unit i are calculated as:

$$C_i^{\text{th}}(P_i^{\text{th}}) = a_i^{\text{th}}(P^{\text{th}})_i^2 + b_i^{\text{th}}P_i^{\text{th}} + c_i^{\text{th}} \quad i \in \Omega_{\text{th}} \tag{3.1}$$

where a_i^{th}, b_i^{th}, and c_i^{th} are the fuel cost coefficients of the thermal unit i. The total fuel cost is calculated as follows:

$$\text{TC} = \sum_{i \in \Omega_{\text{th}}} C_i^{\text{th}}(P_i^{\text{th}}) \tag{3.2}$$

The operating limits are defined as follows:

$$P_i^{\text{th,min}} \le P_i^{\text{th}} \le P_i^{\text{th,max}} \quad i \in \Omega_{\text{th}} \tag{3.3}$$

© Springer International Publishing AG 2017
A. Soroudi, *Power System Optimization Modeling in GAMS*,
DOI 10.1007/978-3-319-62350-4_3

Table 3.1 The economic dispatch data for five thermal units example

i	a_i^{th} ($\$/MW^2$)	b_i^{th} ($\$/MW$)	c_i^{th} ($\$$)	d_i^{th} (kg/MW^2)	e_i^{th} (kg/MW)	f_i^{th} (kg)	$P_i^{th,min}$ (MW)	$P_i^{th,max}$ (MW)
$g1$	3	20	100	2	−5	3	28	206
$g2$	4.05	18.07	98.87	3.82	−4.24	6.09	90	284
$g3$	4.05	15.55	104.26	5.01	−2.15	5.69	68.00	189.00
$g4$	3.99	19.21	107.21	1.10	−3.99	6.20	76.00	266.00
$g5$	3.88	26.18	95.31	3.55	−6.88	5.57	19.00	53.00

where $P_i^{th,max/min}$ are the maximum/minimum power outputs of thermal unit i. The overall thermal unit economic dispatch is formulated as follows:

$$\min_{P_i^{th}} TC = \sum_{i \in \Omega_{th}} C_i^{th}(P_i^{th}) \tag{3.4a}$$

$$C_i^{th}(P_i^{th}) = a_i^{th}(P_i^{th})^2 + b_i^{th}P_i^{th} + c_i^{th} \quad i \in \Omega_{th} \tag{3.4b}$$

$$P_i^{th,min} \le P_i^{th} \le P_i^{th,max} \quad i \in \Omega_{th} \tag{3.4c}$$

$$\sum_{i \in \Omega_{th}} P_i^{th} \ge L_e \tag{3.4d}$$

The economic dispatch data for five units example is given in Table 3.1. This table has nine columns. The first column is showing the generating unit index. The next three columns indicate the cost coefficients for these thermal units $(a_i^{th}, b_i^{th}, c_i^{th})$. The next three columns describe the emission coefficients for these thermal units $(d_i^{th}, e_i^{th}, f_i^{th})$. The last two columns give the minimum and maximum generating limits of each unit if it is on.

The GAMS code for solving the economic dispatch example (3.4) is given in GCode 3.1.

The solution obtained by Code provides information regarding the defined variables. For example, the generating unit schedules are given in Table 3.2.

As it can be seen in Table 3.2, four attributes of variable P(gen) are given. The *.lo* and *.up* values are already known and are given as the input data in Table 3.1. There are two important information regarding the variable P(gen), namely *.l* and *.m*. The level column shows the optimal values of variables obtained by GAMS. The marginal value (*.m*) is zero (EPS or epsilon) for unit $g1$, $g3$, and $g4$. This is because these generating units are not binding variables (not reached to their limits). On the other hand, $g2$ has a marginal value equal to 110.005. This means that if the value of this variable is increased by very small quantity $(\Delta P(g2))$ then the increase in objective function will be approximately equal to $110.005\Delta P(g2)$. This also holds for generating unit $g5$. The increase of $\Delta P(g5)$ will approximately decrease the objective function equal to $199.605\Delta P(g5)$. This is why $g2$ is at its min limit and $g5$ is at its max limit. The total operating costs would be OF = $\$1.3146 \times 10^5$.

GCode 3.1 Economic dispatch problem for five units, example (3.4)

```
set  Gen  /g1*g5/;
scalar  load  /400/;
Table  data(Gen,*)
          a      b      c        d      e       f     Pmin  Pmax
g1        3      20     100      2      -5      3     28    206
g2        4.05   18.07  98.87    3.82   -4.24   6.09  90    284
g3        4.05   15.55  104.26   5.01   -2.15   5.69  68    189
g4        3.99   19.21  107.21   1.1    -3.99   6.2   76    266
g5        3.88   26.18  95.31    3.55   -6.88   5.57  19          53;
variables  P(gen),OF;
equations
eq1,eq2;
eq1  .. OF=e=sum(gen,data(gen,'a')*P(gen)*P(gen)+data(gen,'b')*P
    (gen)+data(gen,'c'));
eq2  .. sum(gen,P(gen))=g=load;
P.lo(gen)=data(gen,'Pmin');
P.up(gen)=data(gen,'Pmax');
model  ECD  /eq1,eq2/;
solve  ECD  us  qcp  min  of;
```

Table 3.2 The ED solution for five units example

P_i^{th}	Lower	Level	Upper	Marginal
g1	28	102.844	206	EPS
g2	90	90	284	110.005
g3	68	76.73	189	–
g4	76	77.425	266	EPS
g5	19	53	53	−199.605

Sometimes we need to perform some sensitivity analysis. For example, we might need to know how the dispatching pattern would change if the load value changes. The load is defined as a scalar quantity in the developed GCode 3.1. The problem is feasible for any load value in the interval [281, 998]. The boundary values of this interval are obtained by the sum of min and max values of generating units given in Table 3.1.

The solution for load sensitivity analysis in ED obtained by GCode 3.2 is given in Table 3.3.

The results of load sensitivity analysis are shown in Fig. 3.1.

3.2 Thermal Unit Environmental Dispatch

The environmental dispatch is referred to the dispatch of electric power plants considering environmental concerns. The environmental pollution produced by power plants (thermal) depends on the fuel type they use to produce electricity

GCode 3.2 Example of sensitivity analysis regarding load value in ED

```
Sets  Gen          /g1*g5/
      counter  /c1*c11/;
Parameter  report(counter,*),repGen(counter,Gen);
Scalar  load  /400/;
Table  data(Gen,*)
         a      b        c       d      e       f     Pmin  Pmax
g1       3      20       100     2      −5      3     28    206
g2       4.05   18.07    98.87   3.82   −4.24   6.09  90    284
g3       4.05   15.55    104.26  5.01   −2.15   5.69  68    189
g4       3.99   19.21    107.21  1.1    −3.99   6.2   76    266
g5       3.88   26.18    95.31   3.55   −6.88   5.57  19    53;
Variables  P(gen),OF;
Equations  eq1,eq2;
eq1  .. OF=e=sum(gen,data(gen,'a')*P(gen)*P(gen)+data(gen,'b')*P
    (gen)+data(gen,'c'));
eq2  .. sum(gen,P(gen))=g=load;
P.lo(gen)=data(gen,'Pmin');
P.up(gen)=data(gen,'Pmax');
Model  ECD  /eq1,eq2/;
loop(counter,
load=sum(gen,data(gen,'Pmin'))
+((ord(counter)−1)/(card(counter)−1))*sum(gen,data(gen,'Pmax')−
    data(gen,'Pmin'));
Solve  ECD  us  qcp  min  of;
repGen(counter,Gen)=P.l(gen);
report(counter,'OF')=of.l;
report(counter,'load')=load;
);
display  repgen,report;
```

Table 3.3 The ED solution for load sensitivity analysis

Iteration	g1	g2	g3	g4	g5	L_e (MW)	OF ($/h)
c1	28.00	90.00	68.00	76.00	19.00	281.00	84,037.85
c2	67.39	90.00	68.00	76.00	51.31	352.70	105,758.02
c3	112.63	90.00	83.98	84.79	53.00	424.40	147,715.94
c4	136.77	101.55	101.86	102.93	53.00	496.10	203,091.66
c5	158.94	117.97	118.28	119.60	53.00	567.80	268,132.18
c6	181.12	134.40	134.71	136.28	53.00	639.50	342,712.41
c7	203.29	150.82	151.14	152.95	53.00	711.20	426,832.36
c8	206.00	173.71	174.02	176.18	53.00	782.90	522,137.81
c9	206.00	201.85	189.00	204.75	53.00	854.60	631,691.67
c10	206.00	237.44	189.00	240.86	53.00	926.30	760,550.50
c11	206.00	284.00	189.00	266.00	53.00	998.00	911,044.09

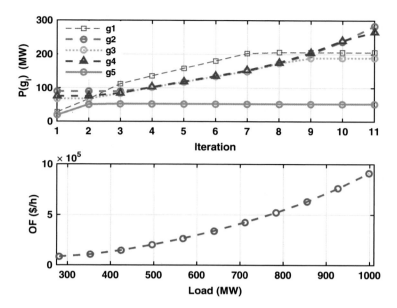

Fig. 3.1 Load sensitivity analysis and impact on costs and generation schedules

and also their generation level. The production costs of ith thermal unit are usually defined as:

$$\text{TE}_i^{th}(P_i^{th}) = d_i^{th}(P_i^{th})^2 + e_i^{th}P_i^{th} + f_i^{th} \quad i \in \Omega_{th} \tag{3.5}$$

where d_i^{th}, e_i^{th}, and f_i^{th} are the fuel emission coefficients of the thermal unit i. TE_i^{th} is the total emission produced by thermal unit i and is expressed as (kg/h).

Different strategies have been proposed to take care of environmental issues in dispatching the power plants, and some of them are listed below:

- Ignoring the environmental pollution and just minimizing the total fuel costs (ED). Optimization problem described in (3.4) should be solved.
- Minimizing the total environmental pollution as the main objective function (END). In this case, optimization problem described in (3.6) should be solved.

$$\min_{P_i^{th}} \text{TE} \tag{3.6a}$$

$$\text{TC} = \sum_{i \in \Omega_{th}} a_i^{th}(P_i^{th})^2 + b_i^{th}P_i^{th} + c_i^{th} \tag{3.6b}$$

$$\text{TE} = \sum_{i \in \Omega_{th}} d_i^{th}(P_i^{th})^2 + e_i^{th}P_i^{th} + f_i^{th} \tag{3.6c}$$

$$P_i^{\text{th,min}} \le P_i^{\text{th}} \le P_i^{\text{th,max}} \quad i \in \Omega_{\text{th}} \tag{3.6d}$$

$$\sum_{i \in \Omega_{\text{th}}} P_i^{\text{th}} \ge L_e \tag{3.6e}$$

The solution of (3.6) can be found in the first column of Table 3.4.
- Adding the environmental pollution is a penalty to the cost function (penalty). For this strategy, the environmental pollution price (C_e) should be known. In this case, optimization problem described in (3.7) should be solved.

$$\min_{P_i^{\text{th}}} \text{OF} = \text{TC} + C_e \times \text{TE} \tag{3.7a}$$

$$\text{TC} = \sum_{i \in \Omega_{\text{th}}} a_i^{\text{th}} (P_i^{\text{th}})^2 + b_i^{\text{th}} P_i^{\text{th}} + c_i^{\text{th}} \tag{3.7b}$$

$$\text{TE} = \sum_{i \in \Omega_{\text{th}}} d_i^{\text{th}} (P_i^{\text{th}})^2 + e_i^{\text{th}} P_i^{\text{th}} + f_i^{\text{th}} \tag{3.7c}$$

$$P_i^{\text{th,min}} \le P_i^{\text{th}} \le P_i^{\text{th,max}} \quad i \in \Omega_{\text{th}} \tag{3.7d}$$

$$\sum_{i \in \Omega_{\text{th}}} P_i^{\text{th}} \ge L_e \tag{3.7e}$$

The solution of (3.7) can be found in the second column of Table 3.4.
- Adding the environmental pollution as a constraint to the original economic dispatch problem (Elimit). In this case, optimization problem described in (3.8) should be solved.

$$\min_{P_i^{\text{th}}} \text{OF} = \text{TC} \tag{3.8a}$$

$$\text{TC} = \sum_{i \in \Omega_{\text{th}}} a_i^{\text{th}} (P_i^{\text{th}})^2 + b_i^{\text{th}} P_i^{\text{th}} + c_i^{\text{th}} \tag{3.8b}$$

$$\text{TE} = \sum_{i \in \Omega_{\text{th}}} d_i^{\text{th}} (P_i^{\text{th}})^2 + e_i^{\text{th}} P_i^{\text{th}} + f_i^{\text{th}} \tag{3.8c}$$

$$P_i^{\text{th,min}} \le P_i^{\text{th}} \le P_i^{\text{th,max}} \quad i \in \Omega_{\text{th}} \tag{3.8d}$$

$$\sum_{i \in \Omega_{\text{th}}} P_i^{\text{th}} \ge L_e \tag{3.8e}$$

$$\text{TE} \le \text{Elimit} \tag{3.8f}$$

The solution of (3.8) can be found in the third column of Table 3.4.
- Solving the dispatch problem as multi-objective problem. In this case, optimization problem described in (3.9) should be solved.

$$\min_{P_i^{\text{th}}} \text{TC} = \sum_{i \in \Omega_{\text{th}}} a_i^{\text{th}} (P_i^{\text{th}})^2 + b_i^{\text{th}} P_i^{\text{th}} + c_i^{\text{th}} \tag{3.9a}$$

$$\min_{P_i^{\text{th}}} \text{TE} = \sum_{i \in \Omega_{\text{th}}} d_i^{\text{th}} (P_i^{\text{th}})^2 + e_i^{\text{th}} P_i^{\text{th}} + f_i^{\text{th}} \tag{3.9b}$$

$$P_i^{\text{th,min}} \leq P_i^{\text{th}} \leq P_i^{\text{th,max}} \quad i \in \Omega_{\text{th}} \tag{3.9c}$$

$$\sum_{i \in \Omega_{\text{th}}} P_i^{\text{th}} \geq L_e \tag{3.9d}$$

The ED, END, Penalty, and Limit strategies are solved using GCode 3.3. In this code, you can find four solve statements.

- ED strategy: The first solve statement minimizes the total costs without any concern about environmental issues.
- END strategy: The second solve statement minimizes the total environmental pollution without any concern about economic issues.
- Penalty strategy: The third solve statement minimizes the total fuel costs + environmental pollution costs.
- Limited Emission strategy: The fourth solve statement minimizes the total fuel costs while environmental pollution is limited to some value.

Each solve statement obtains the optimal generating schedules in each strategy. A parameter $(report(gen, *))$ is defined which is updated after each solve statement.

The solution for different environmental dispatching strategies is given in Table 3.4.

In some occasions, it is needed to solve the problem in a multi-objective way. The multi-objective model is described in (3.9) and coded in GCode 3.4.

The Pareto optimal front in multi-objective environmental dispatch strategy for load = 400 MW is depicted in Fig. 3.2. The operating schedules of different units in solutions of Pareto optimal front are plotted in Fig. 3.3 for load = 400 MW.

The Pareto optimal front found in Fig. 3.2 depends on technical characteristics of generating units in one hand and also on load value (= 400 MW). The question is what happens if the load increases. How can we obtain the Pareto optimal front for different values of loads? In order to obtain each Pareto optimal front, it is needed to

Table 3.4 Solution for different environmental dispatching strategies

P(g)	ED	END	Penalty	Elimit
g1	102.84	71.62	103.05	94.21
g2	90.00	90.00	90.00	90.00
g3	76.73	68.00	75.18	68.00
g4	77.43	129.76	78.77	94.79
g5	53.00	40.62	53.00	53.00
OF ($)	135,310.00	152,170.00	140,990.00	142,190.00
TE (kg)	96,450.75	87,089.40	94,428.55	90,000.00
TC ($)	131,460.00	148,680.00	131,550.00	133,190.00

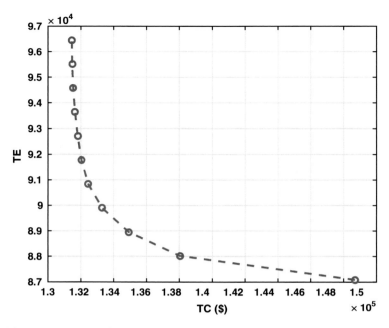

Fig. 3.2 Pareto optimal front in multi-objective environmental dispatch strategy for load = 400 MW

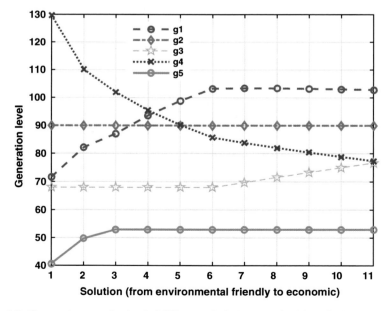

Fig. 3.3 Changes in generating level of different units in Pareto optimal front for load = 400 MW

GCode 3.3 Example of different environmental dispatch strategies

```
Set  Gen  / g1*g5 /;
Parameter  report(gen,*);
Scalars
load  /400/
Eprice  /0.1/
Elim  /90000/;
Table  data(Gen,*)
         a      b      c        d      e      f      Pmin  Pmax
g1       3      20     100      2      −5     3      28    206
g2       4.05   18.07  98.87    3.82   −4.24  6.09   90    284
g3       4.05   15.55  104.26   5.01   −2.15  5.69   68    189
g4       3.99   19.21  107.21   1.1    −3.99  6.2    76    266
g5       3.88   26.18  95.31    3.55   −6.88  5.57   19    53;
variables  P(gen),OF,TE,TC;
equations
eq1,eq2,eq3,eq4;
eq1  ..  TC=e=sum(gen,data(gen,'a')*P(gen)*P(gen)+data(gen,'b')*P
    (gen)+data(gen,'c'));
eq2  ..  sum(gen,P(gen))=g=load;
eq3  ..  TE=e=sum(gen,data(gen,'d')*P(gen)*P(gen)+data(gen,'e')*P
    (gen)+data(gen,'f'));
eq4  ..  OF=e=TC+TE*Eprice;
P.lo(gen)=data(gen,'Pmin');
P.up(gen)=data(gen,'Pmax');
Model  END  /eq1,eq2,eq3,eq4/;
Solve  END  us  qcp  min  TC;
report(gen,'ED')=P.l(gen);
Solve  END  us  qcp  min  te;
report(gen,'END')=P.l(gen);
Solve  END  us  qcp  min  OF;
report(gen,'penalty')=P.l(gen);
TE.up=Elim;
Solve  END  us  qcp  min  TC;
report(gen,'limit')=P.l(gen);
Display  report;
```

have a loop. The ϵ constraint method [1] is used to obtain the Pareto optimal front. The GCode 3.5 finds the Pareto optimal fronts for different load values 400, 450, 500, and 550 MW. It uses nested loops to change the load value in upper level loop and changes the epsilon value in lower level loop.

The Pareto optimal fronts for different load values are depicted in Fig. 3.4.

3.3 CHP Economic Dispatch

The combined heat and power (CHP) technology can generate heat and electric power simultaneously. In CHP-related dispatch problems, we are usually faced with different sources of energy to supply heat and power demand. These sources of

GCode 3.4 Example of multi-objective environmental dispatch strategy

```
Set  Gen  /g1*g5/
     counter  /c1*c11/;
parameter  report(*),rep(counter,*);
Scalars
load  /400/
Eprice  /0.1/
Elim;
Table  data(Gen,*)
        a     b      c        d      e      f     Pmin  Pmax
g1      3     20     100      2      -5     3     28    206
g2      4.05  18.07  98.87    3.82   -4.24  6.09  90    284
g3      4.05  15.55  104.26   5.01   -2.15  5.69  68    189
g4      3.99  19.21  107.21   1.1    -3.99  6.2   76    266
g5      3.88  26.18  95.31    3.55   -6.88  5.57  19    53;
Variables  P(gen),OF,TE,TC;
Equations
eq1,eq2,eq3;
eq1  ..  TC=e=sum(gen,data(gen,'a')*P(gen)*P(gen)+data(gen,'b')*P
     (gen)+data(gen,'c'));
eq2  ..  sum(gen,P(gen))=g=load;
eq3  ..  TE=e=sum(gen,data(gen,'d')*P(gen)*P(gen)+data(gen,'e')*P
     (gen)+data(gen,'f'));
P.lo(gen)=data(gen,'Pmin');
P.up(gen)=data(gen,'Pmax');
model END /eq1,eq2,eq3/;
solve END us qcp min TC;
report('maxTE')=TE.l;
report('minTC')=TC.l;
Solve END us qcp min te;
report('maxTC')=TC.l;
report('minTE')=TE.l;
loop(counter,
Elim=(report('maxTE')-report('minTE'))*((ord(counter)-1)/(card
(counter)-1))+report('minTE');
TE.up=Elim;
Solve END us qcp min TC;
rep(counter,'TC')=TC.l;
rep(counter,'TE')=TE.l;
);
display rep;
```

GCode 3.5 Sensitivity analysis for load values in multi-objective environmental dispatch strategy

```
sets
Gen  /g1*g5/
counter  /c1*c11/
Loadcounter  /Lc1*Lc4/;
parameter  report(Loadcounter,*),rep(Loadcounter,counter,*),
rep2(Loadcounter,counter,gen);
scalars
load  /400/
Eprice  /0.1/
Elim;
Table  data(Gen,*)
        a       b       c       d       e       f       Pmin  Pmax
g1      3       20      100     2       −5      3       28     206
g2      4.05    18.07   98.87   3.82    −4.24   6.09    90     284
g3      4.05    15.55   104.26  5.01    −2.15   5.69    68     189
g4      3.99    19.21   107.21  1.1     −3.99   6.2     76     266
g5      3.88    26.18   95.31   3.55    −6.88   5.57    19     53;
Variables  P(gen),OF,TE,TC;
Equations
eq1,eq2,eq3;
eq1  ..  TC=e=sum(gen,data(gen,'a')*P(gen)*P(gen)+data(gen,'b')*P
    (gen)+data(gen,'c'));
eq2  ..  sum(gen,P(gen))=g=load;
eq3  ..  TE=e=sum(gen,data(gen,'d')*P(gen)*P(gen)+data(gen,'e')*P
    (gen)+data(gen,'f'));
P.lo(gen)=data(gen,'Pmin');
P.up(gen)=data(gen,'Pmax');
Model  END  /eq1,eq2,eq3/;
Loop(Loadcounter,
load=350+ord(Loadcounter)*50;
solve END us qcp min TC;
report(Loadcounter,'maxTE')=TE.1;
report(Loadcounter,'minTC')=TC.1;
Solve END us qcp min te;
report(Loadcounter,'maxTC')=TC.1;
report(Loadcounter,'minTE')=TE.1;

loop(counter,
Elim=(report(Loadcounter,'maxTE')
−report(Loadcounter,'minTE'))*((ord(counter)−1)/(card(counter)
    −1))
+report(Loadcounter,'minTE');
TE.up=Elim;
solve END us qcp min TC;
rep(Loadcounter,counter,'TC')=TC.1;
rep(Loadcounter,counter,'TE')=TE.1;
rep2(Loadcounter,counter,gen)=P.1(gen);
);
TE.up=inf;
);
Display rep,rep2,report;
```

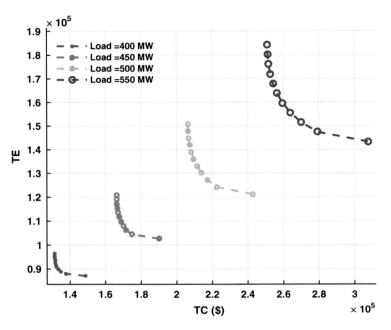

Fig. 3.4 Pareto optimal fronts for different load values

energy are usually thermal plants, CHP units, and heat only units. As described in Sect. 3.1, the fuel cost of a thermal unit depends on its electricity generation level. Similarly, the operating costs of CHP units depend on the level of heat and electricity they generate. Additionally, CHP units have a feasible operating region that dictates some limits for power and heat that they can generate. An approximate heat-power operating region for a CHP unit is shown in Fig. 3.5 [2].

The CHP economic dispatch problem can be modeled as described in (3.10):

$$\min_{P_i^{\text{th}}, P_j^{\text{chp}}, q_j^{\text{chp}}, q_k^h} \text{OF} = F^{\text{th}} + F^h + F^{\text{chp}} \tag{3.10a}$$

$$F^{\text{th}} = \sum_{i \in \Omega_{\text{th}}} a_i^{\text{th}}(P_i^{\text{th}})^2 + b_i^{\text{th}} P_i^{\text{th}} + c_i^{\text{th}} \tag{3.10b}$$

$$F^h = \sum_{k \in \Omega_h} a_k^h(q_k^h)^2 + b_k^h q_k^h + c_k^h \tag{3.10c}$$

$$F^{\text{chp}} = \sum_{j \in \Omega_{\text{chp}}} a_j^{\text{chp}}(P_j^{\text{chp}})^2 + b_j^{\text{chp}} P_j^{\text{chp}} + c_j^{\text{chp}} \tag{3.10d}$$

$$+ d_j^{\text{chp}}(q_j^{\text{chp}})^2 + e_j^{\text{chp}} q_j^{\text{chp}} + f_j^{\text{chp}} P_j^{\text{chp}} q_j^{\text{chp}}$$

$$P_i^{\text{th,min}} \leq P_i^{\text{th}} \leq P_i^{\text{th,max}} \quad i \in \Omega_{\text{th}} \tag{3.10e}$$

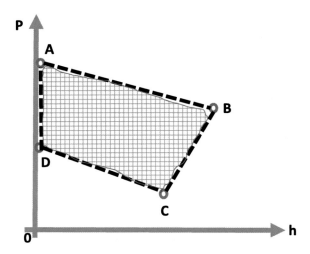

Fig. 3.5 Approximate heat-power operating region for a CHP unit

$$Q_k^{h,\min} \leq q_k^h \leq Q_k^{h,\max} \quad k \in \Omega_h \tag{3.10f}$$

$$P_j^{chp,\min}(q_j^{chp}) \leq P_j^{chp} \leq P_j^{chp,\max}(q_j^{chp}) \quad j \in \Omega_{chp} \tag{3.10g}$$

$$Q_j^{chp,\min}(P_j^{chp}) \leq q_j^{chp} \leq Q_j^{chp,\max}(P_j^{chp}) \quad j \in \Omega_{chp} \tag{3.10h}$$

$$\sum_{i \in \Omega_{th}} P_i^{th} + \sum_{j \in \Omega_{chp}} P_j^{chp} \geq L_e \tag{3.10i}$$

$$\sum_{k \in \Omega_h} q_k^h + \sum_{j \in \Omega_{chp}} q_j^{chp} \geq L_h \tag{3.10j}$$

The objective function defined in (3.10a) has three terms namely thermal unit (F^{th}) [calculated in (3.10b)], heat only unit (F^h) [calculated in (3.10c)], and CHP unit (F^{chp}) costs [calculated in (3.10d)]. The decision variables are electric power generation of thermal units (P_i^{th}) and CHP units P_j^{chp} and also the heat generated by CHP units q_j^{chp} and heat only units q_k^h as described in (3.10a). The operating limits of thermal units and heat only units are modeled in (3.10e) and (3.10f), respectively as given in Table 3.5. Unlike the other technologies mentioned before, the operating limits of CHP units are dependent on electricity and heat generation level as described in (3.10g) and (3.10h) as described in Table 3.6. The electricity and heat supply-demand balance are modeled in (3.10i) and (3.10j), respectively.

The electric and heat demands are assumed to be 605 MW and 540 MW, respectively.

Table 3.5 Characteristic data of different technologies

Technology	a	b	c	d	e	f	p^{min}	p^{max}	h^{min}	h^{max}
g1	3.00	20.00	100.00				30.00	180.00		
g2	4.05	18.07	98.87				90.00	290.00		
h1	4.05	10.55	104.26						60.00	200.00
h2	3.99	9.21	107.21						70.00	270.00
chp1	0.0345	14.00	2540.00	0.03	4.20	0.031				
chp2	0.0435	13.00	1460.00	0.02	0.70	0.011				

Table 3.6 Feasible operating regions of CHP units

CHP unit	Aq	Ap	Bq	Bp	Cq	Cp	Dq	Dp
chp1	0	247	180	215	104.8	81	0	99
chp2	0	125	135	110	75	40	0	45

GCode 3.6 CHP dispatching GAMS code for Example (3.10)

```
Sets
Gen /g1*g2/
heat /h1*h2/
CHP /chp1*chp2/;
scalars
Le /605/
Lh /540/;
Table dataTh(Gen,*)
      a       b       c       d      e      f      Pmin  Pmax  hmin  hmax
g1    3       20      100     0      0      0      28    206   0     0
g2    4.05    18.07   98.87   0      0      0      90    284   0     0 ;
Table dataH(heat,*)
      a       b       c       d      e      f      Pmin  Pmax  hmin  hmax
h1    4.05    10.55   104.26  0      0      0      0     0     60    200
h2    3.99    9.21    107.21  0      0      0      0     0     70    270;
Table datachp(chp,*)
      a       b       c       d      e      f      Pmin  Pmax  hmin  hmax
chp1  0.0345  14      2540    0.03   4.2    0.031  0     0     0     0
chp2  0.0435  13      1460    0.02   0.7    0.011  0     0     0     0;

Table FR(chp,*)
      Aq      Ap      Bq      Bp     Cq     Cp     Dq    Dp
chp1  0       247     180     215    104.8  81     0     99
chp2  0       125     135     110    75     40     0     45;

Variables  P(gen),OF,q(heat), pchp(chp), qchp(chp),Fth,Fh,Fchp;
Equations  eq1,eq2,eq3,eq4,eq5,eq6,eq7a,eq7b,eq7c;
eq1  ..  OF=e=Fth+Fh+Fchp;
eq2  ..  Fth=e=sum(gen,dataTh(gen,'a')*P(gen)*P(gen)+dataTh(gen,'
         b')*P(gen)+dataTh(gen,'c'));
eq3  ..  Fh=e=sum(heat,dataH(heat,'a')*q(heat)*q(heat)+dataH(heat
         ,'b')*q(heat)+dataH(heat,'c'));
eq4 .. Fchp=e=sum(CHP,datachp(CHP,'a')*Pchp(CHP)*Pchp(CHP)
+datachp(CHP,'b')*Pchp(CHP)+datachp(CHP,'c'))+
```

```
sum(CHP,datachp(CHP,'d')*qchp(CHP)*qchp(CHP)
+datachp(CHP,'e')*qchp(CHP)+datachp(CHP,'f')*qchp(CHP)
    ));
eq5 .. sum(gen,P(gen))+sum(CHP,Pchp(CHP))=g=Le;
eq6 .. sum(heat,q(heat))+sum(CHP,Qchp(CHP))=g=Lh;

eq7a(CHP) ..    Pchp(CHP)-FR(chp,'Dp')=g=
(qchp(CHP)-FR(chp,'Dq'))*(FR(chp,'Dp')-FR(chp,'Cp'))/(FR(chp,'
    Dq')-FR(chp,'Cq'));
eq7b(CHP) ..    Pchp(CHP)-FR(chp,'Ap')=l=
(qchp(CHP)-FR(chp,'Dq'))*(FR(chp,'Ap')-FR(chp,'Bp'))/(FR(chp,'
    Aq')-FR(chp,'Bq'));
eq7c(CHP) ..    Pchp(CHP)-FR(chp,'Bp')=g=
(qchp(CHP)-FR(chp,'Bq'))*(FR(chp,'Bp')-FR(chp,'Cp'))/(FR(chp,'
    Bq')-FR(chp,'Cq'));
P.lo(gen)=dataTh(gen,'Pmin');
P.up(gen)=dataTh(gen,'Pmax');
q.lo(heat)=dataH(heat,'hmin');
q.up(heat)=dataH(heat,'hmax');
Model chpdispatch /all/;
Solve chpdispatch us nlp min OF;
```

Table 3.7 Solution of dispatching CHP units

Technology	p	Technology	q
chp1	215	chp1	180
chp2	110	chp2	135
g1	160.714	h1	111.577
g2	119.286	h2	113.423
L_e	605	L_h	540

The total operating costs are $F^{th} = 1.4068 \times 10^5$ ($/h), $F^h = 1.0418 \times 10^5$ ($/h), $F^{chp}=14,111.162$ ($/h) (Table 3.7).

3.4 Hydro Unit Economic Dispatch

This section addresses the optimal operation of a hydro generating power plant. This power plant comprises several cascaded plants along a river basin. The objective is to minimize the total operating costs while considering different technical constraints such as relationship between the generated power, water discharged, and the head of the associated reservoir. The concept of cascaded hydro units is shown in Fig. 3.6. The water balance equations that should be satisfied in each hour are as follows:

$$2L_{t+1}^h = L_t^h + I_{t+1}^h - R_{t+1}^h - S_{t+1}^h \tag{3.11}$$

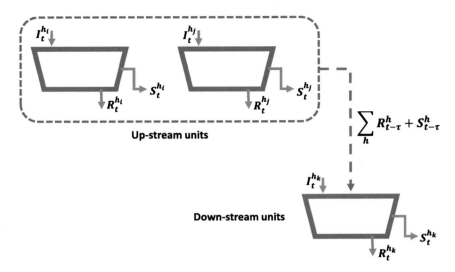

Fig. 3.6 Cascaded hydro units

$$+\eta \sum_{\hat{h}} \left[R^{\hat{h}}_{t+1-\tau_{\hat{h}}} + S^{\hat{h}}_{t+1-\tau_{\hat{h}}} \right]$$

$$L^h_{\min} \leq L^h_t \leq L^h_{\max}, \hat{h} \in \text{up}\{h\} \tag{3.12}$$

$$R^h_t \leq R^h_{\max}, L^h_{t0} = L^h_{\text{ini}}, L^h_{t24} = L^h_{\text{fin}} \tag{3.13}$$

where L^h_t is reservoir volume, I^h_{t+1} is the water inflow, R^h_t is the released water, and S^h_t is the spilled water at the end of period t in million m^3. $R_{\max}$ is the maximum released capacity per hour in million m^3. L^h_{ini}, L^h_{fin} are the volume of the water in dam at beginning and end of the considered horizon, respectively. This constraint means that the volume of water in a reservoir of hydro turbine h in time $t+1$ will be equal to its value in the previous period plus the water inflow to its reservoir in time $t+1$ minus its own released/spilled water and in time $t+1$ plus the released/spilled water of all reservoirs in its upstream in previous hours (with considering time delays $\tau_{\hat{h}}$). The concept of this cascade reservoir water balance constraint is depicted in Fig. 3.6.

The hydro power production function (HPF) (or hill chart [3]) which relates the output power of hydro plant to the water level, inflow, and spillage is of great importance in hydro plant scheduling. In this chapter, the method proposed in [4] has been adopted which describes the relationship between the released water and water head of the reservoir with the output power of the hydro power plant, as follows:

$$P^h_t = c^h_1(L^h_t)^2 + c^h_2(R^h_t)^2 + c^h_3 R^h_t L^h_t + c^h_4 L^h_t + c^h_5 R^h_t + c^h_6 \tag{3.14}$$

Table 3.8 Techno-economic characteristics of thermal units

Unit	a_i^{th} ($/MW^2)	b_i^{th} ($/MW)	c_i^{th} ($)	$P_i^{th,min}$ (MW)	$P_i^{th,max}$ (MW)	RU$_i$ (MW)	RD$_i$ (MW)
g1	0.00043	16.6	900	100	400	60	60
g2	0.00073	15.5	800	130	400	40	40
g3	0.00059	14.8	700	70	300	60	60
g4	0.00075	15.9	470	60	300	30	30
g5	0.00079	16.6	200	80	250	50	50

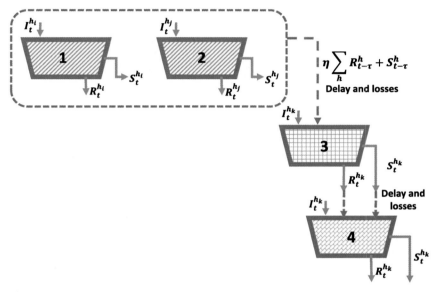

Fig. 3.7 Four cascaded hydro units configuration

Equation (3.14) states that the generated power in time t is nonlinearly related to the water head (L_t^h) and released water (R_t^h) in time t simultaneously. The coefficients $c_{1\rightarrow6}^h$ are the characteristic factors of hydro turbine h which describe this relation. Now we can solve a mixed hydro-thermal economic dispatch problem. Suppose we have five thermal units which their techno-economic characteristics are described in Table 3.8. Additionally, there are four cascaded hydro power units as shown in Fig. 3.7 which their techno-economic characteristics are described in Table 3.9. The loss coefficient η is assumed to be 0.9 (means 10% of released/spilled water vaporizes). The operating cost of hydro power plant is assumed to be constant $\zeta_h = 14\$/MW$.

The hourly water inflows for each hydro power and hourly electric demand are given in Table 3.10.

The hydro-thermal economic dispatch problem is formulated in (3.15).

Table 3.9 Techno-economic characteristics of hydro units

Unit	c_1^h	c_2^h	c_3^h	c_4^h	c_5^h	c_6^h	L_{min}^h	L_{max}^h	L_{ini}^h	L_{fin}^h	R_{min}^h	R_{min}^h	$P^{h,min}$	$P^{h,max}$	τ
$h1$	−0.0042	−0.42	0.030	0.90	10	−50	80	150	100	120	5	15	0	500	2
$h2$	−0.0040	−0.30	0.015	1.14	9.5	−70	60	120	80	70	6	15	0	500	1
$h3$	−0.0016	−0.30	0.014	0.55	5.5	−40	100	240	170	170	10	30	0	500	4
$h4$	−0.0030	−0.31	0.027	1.44	14	−90	70	160	120	140	6	20	0	500	0

Table 3.10 Hourly water inflows for each hydro power and hourly electric demand

Time	I_t^{h1}	I_t^{h2}	I_t^{h3}	I_t^{h4}	$L_{e,t}$ (MW)
t_1	10	8	8.1	2.8	1275
t_2	9	8	8.2	2.4	1326
t_3	8	9	4	1.6	1190
t_4	7	9	2	0	1105
t_5	6	8	3	0	1139
t_6	7	7	4	0	1360
t_7	8	6	3	0	1615
t_8	9	7	2	0	1717
t_9	10	8	1	0	1853
t_{10}	11	9	1	0	1836
t_{11}	12	9	1	0	1870
t_{12}	10	8	2	0	1955
t_{13}	11	8	4	0	1887
t_{14}	12	9	3	0	1751
t_{15}	11	9	3	0	1717
t_{16}	10	8	2	0	1802
t_{17}	9	7	2	0	1785
t_{18}	8	6	2	0	1904
t_{19}	7	7	1	0	1819
t_{20}	6	8	1	0	1785
t_{21}	7	9	2	0	1547
t_{22}	8	9	2	0	1462
t_{23}	9	8	1	0	1445
t_{24}	10	8	0	0	1360

$$\min_{DV} OF = \sum_t \left(P_t^h \zeta_h + \sum_i \left(a_i^{th}(P_{i,t}^{th})^2 + b_i^{th} P_{i,t}^{th} + c_i^{th} \right) \right) \tag{3.15a}$$

$$L_{t+1}^h = L_t^h + I_{t+1}^h - R_{t+1}^h - S_{t+1}^h + \eta \sum_{\hat{h}} [R_{t+1-\tau_{\hat{h}}}^{\hat{h}} + S_{t+1-\tau_{\hat{h}}}^{\hat{h}}] \tag{3.15b}$$

$$L_{min}^h \leq L_t^h \leq L_{max}^h, \hat{h} \in \text{up} \{h\} \tag{3.15c}$$

$$R_{min}^h \leq R_t^h \leq R_{max}^h \tag{3.15d}$$

$$L_{t_0}^h = L_{\text{ini}}^h, L_{t_{24}}^h = L_{\text{fin}}^h \tag{3.15e}$$

$$P_t^h = c_1^h(L_t^h)^2 + c_2^h(R_t^h)^2 + c_3^h R_t^h L_t^h + c_4^h L_t^h + c_5^h R_t^h + c_6^h \tag{3.15f}$$

$$\sum_h P_{h,t} + \sum_i P_{i,t}^{\text{th}} \geq L_{e,t} \tag{3.15g}$$

$$P_{i,t}^{\text{th}} - P_{i,t-1}^{\text{th}} \leq \text{RU}_i \tag{3.15h}$$

$$P_{i,t-1}^{\text{th}} - P_{i,t}^{\text{th}} \leq \text{RD}_i \tag{3.15i}$$

$$P_i^{\text{th,min}} \leq P_i^{\text{th}} \leq P_i^{\text{th,max}} \quad i \in \Omega_{\text{th}} \tag{3.15j}$$

$$P^{h,\text{min}} \leq P_t^h \leq P^{h,\text{max}} \tag{3.15k}$$

$$\text{DV} \in \{P_{i,t}^{\text{th}}, P_t^h, R_t^h, I_t^h, S_t^h\} \tag{3.15l}$$

The GCode 3.7 is developed to solve the hydro-thermal economic dispatch problem described in (3.15). In this code, for saving the space, table demand and table inflow are shown partially. The optimal hourly dispatch of hydro and thermal units are given in Table 3.11.

GCode 3.7 Hydro-thermal economic dispatch Example (3.15)

```
Sets t    /t1*t24/,  H /h1*h4/,  i /g1*g5/;  Scalar zeta /14/;
Table data(H,*)
     c1        c2    c3     c4     c5     c6
h1  −0.0042  −0.44  0.040  0.80   11.0  −53
h2  −0.0043  −0.32  0.013  1.24   9.7   −71
h3  −0.0015  −0.31  0.012  0.54   5.7   −42
h4  −0.0032  −0.33  0.025  1.43   14.1  −91;
Table  gendata(i,*)  generator  cost  characteristics  and  limits
      a         b      c     lowlim  upplim  RU    RD
g1    0.00043  16.6  900    100     400     60    60
g2    0.00073  15.5  800    130     400     40    40
g3    0.00059  14.8  700    70      300     60    60
g4    0.00075  15.9  470    60      300     30    30
g5    0.00079  16.6  200    80      250     50    50;
Alias(H,Hhat);  set  upstream(H,Hhat);
upstream('h3',Hhat)$(ord(Hhat)<3)=yes;  upstream('h4','h3')=yes;
Parameter  delay(H)
/h1 2
h2   1
h3   4
h4   0/;
Table  inflow(t,H)
       h1             h2           h3           h4
t1     10             8            8.1          2.8
t2     9              8            8.2          2.4
t23    9              8            1            0
t24    10             8            0            0;
parameter  demand(t)
/t1         1275
```

```
t2          1326
t23         1445
t24          1360/;
Table  charac(H,*)
        Vmin Vmax Vini Vfin Qmin Qmax Pmin Pmax
h1      80    150   100  120   5    15    0   500
h2      60    120   80   70    6    15    0   500
h3      100   240   170  170   10   30    0   500
h4      70    160   120  140   6    20    0   500;
Variables  V(H,t),R(H,t),Spill(H,t),OBJ,PH(H,t),costThermal,p(i,
    t);
p.up(i,t) = gendata(i,"upplim") ; p.lo(i,t) = gendata(i,"lowlim
    ");
V.LO(H,t) = charac(H,'Vmin'); V.UP(H,t) = charac(H,'Vmax');
V.FX(H,'t24') = charac(H,'Vfin');
ph.lo(H,t)=charac(H,'Pmin'); ph.up(H,t)=charac(H,'Pmax');;
R.LO(H,t) = charac(H,'Qmin'); R.UP(H,t) = charac(H,'Qmax');
Spill.LO(H,t) = 0;
Equations
Waterlevel ,pcalc,Genconst3,Genconst4,costThermalcalc,balance,
    OFdef;
costThermalcalc.. costThermal=e=sum((t,i),gendata(i,'a')*power
    (p(i,t),2)+gendata(i,'b')*p(i,t)+gendata(i,'c'));
Genconst3(i,t) .. p(i,t+1)-p(i,t)=l=gendata(i,'RU');
Genconst4(i,t) .. p(i,t-1)-p(i,t)=l=gendata(i,'RD');
Waterlevel(H,t+1).. V(H,t+1) =e= charac(H,'Vini')$(ord(t)=1)+
V(H,t)$(ord(t)>1)+inflow(t+1,H)-R(H,t+1)-Spill(H,t+1)+
0.9*sum(Hhat$upstream(H,Hhat),R(Hhat,t-delay(H))+Spill(Hhat,t-
    delay(H)))  ;
pcalc(H,t) .. Ph(H,t)=e=data(H,'c1')*V(H,t)*V(H,t)+data(H,'c2')
    *R(H,t)*R(H,t)
                  +data(H,'c3')*V(H,t)*R(H,t)
                  +data(H,'c4')*V(H,t)+data(H,'c5')*R(H,t
                  )+data(H,'c6');
balance(t) .. sum(i,p(i,t))+sum(H,ph(H,t))=g=demand(t);
OFdef ..        OBJ=e=zeta*sum((H,t),Ph(H,t))+ costThermal;
Model hydro /all/;
Solve hydro us nlp min OBJ;
```

The thermal unit power schedules are shown in Fig. 3.8. The hydro unit power schedules are shown in Fig. 3.9.

Different models and methodologies related to hydro power scheduling have been addressed in the literature such as: Self-scheduling of a hydro producer in a pool-based electricity market [3], dispatchability enhancement of variable Wind Generation by coordination With Pumped-Storage Hydro Units [5], coordinated wind-hydro bidding strategies in day-ahead markets [6], and self-scheduling of hydro-thermal Genco in smart grids [7].

Table 3.11 Optimal hourly dispatch of hydro and thermal units

Time	h_1			h_2			h_3			h_4			Thermal				
	P	V	R	P	V	R	P	V	R	P	V	R	p_1	p_2	p_3	p_4	p_5
t_1	128.50	150.00	15.00	112.78	120.00	15.00	60.60	234.97	13.74	285.88	160.00	20.00	100.00	207.24	240.00	60.00	80.00
t_2	75.99	101.14	7.86	54.84	82.00	6.00	52.35	157.76	12.25	285.88	160.00	20.00	139.70	247.24	300.00	90.00	80.00
t_3	49.59	104.14	5.00	56.64	85.00	6.00	50.55	149.76	12.00	285.88	160.00	20.00	100.00	222.90	284.44	60.00	80.00
t_4	49.82	106.14	5.00	58.36	88.00	6.00	48.09	139.86	11.90	277.27	150.80	20.00	100.00	200.00	224.44	67.01	80.00
t_5	49.93	107.14	5.00	59.47	90.00	6.00	35.43	100.09	11.13	268.03	141.51	20.00	100.00	240.00	209.13	97.01	80.00
t_6	55.38	108.65	5.49	60.01	91.00	6.00	60.57	240.00	13.84	285.88	160.00	20.00	142.02	280.00	269.13	127.01	80.00
t_7	94.17	106.49	10.16	65.31	90.15	6.85	60.60	233.25	13.71	285.88	160.00	20.00	202.02	320.00	300.00	157.01	130.00
t_8	85.19	106.64	8.85	60.09	91.15	6.00	60.40	222.93	13.51	285.88	160.00	20.00	245.04	360.00	300.00	187.01	133.38
t_9	87.32	107.53	9.11	62.63	92.92	6.23	59.85	211.60	13.29	285.88	160.00	20.00	285.12	400.00	300.00	217.01	155.19
t_{10}	88.86	109.28	9.26	67.55	95.13	6.79	58.94	200.28	13.07	285.88	160.00	20.00	251.09	400.00	300.00	247.01	136.67
t_{11}	91.06	111.80	9.48	72.11	96.76	7.37	57.73	189.40	12.86	285.88	160.00	20.00	250.08	400.00	300.00	277.01	136.12
t_{12}	92.43	112.14	9.66	75.87	96.76	7.99	57.08	184.49	12.76	285.88	160.00	20.00	287.34	400.00	300.00	300.00	156.40
t_{13}	98.80	112.52	10.61	82.83	95.33	9.43	56.37	179.63	12.67	285.88	160.00	20.00	229.52	400.00	300.00	300.00	133.60
t_{14}	83.79	116.10	8.42	72.85	96.85	7.48	55.50	174.22	12.57	285.88	160.00	20.00	169.52	400.00	300.00	299.86	83.60
t_{15}	93.21	117.50	9.60	81.81	96.78	9.07	54.66	169.44	12.47	285.88	160.00	20.00	121.44	400.00	300.00	300.00	80.00
t_{16}	93.99	117.80	9.70	83.50	95.19	9.58	53.71	164.38	12.38	285.88	160.00	20.00	181.44	400.00	300.00	300.00	103.49
t_{17}	90.66	117.54	9.26	82.68	92.46	9.73	52.84	160.05	12.29	285.88	160.00	20.00	182.36	400.00	300.00	300.00	90.59
t_{18}	97.29	115.27	10.26	85.50	87.32	11.15	52.38	157.86	12.25	285.88	160.00	20.00	242.36	400.00	300.00	300.00	140.59
t_{19}	93.39	112.49	9.79	83.90	82.78	11.53	50.84	150.95	12.12	285.88	160.00	20.00	199.31	400.00	300.00	297.20	108.49
t_{20}	97.52	107.85	10.63	82.85	78.68	12.10	49.78	146.54	12.03	285.88	160.00	20.00	171.77	400.00	300.00	267.20	130.00
t_{21}	50.17	109.85	5.00	72.91	78.10	9.58	49.07	143.68	11.97	285.88	160.00	20.00	111.77	360.00	300.00	237.20	80.00
t_{22}	50.36	112.85	5.00	67.77	78.63	8.47	50.79	151.27	11.50	285.88	160.00	20.00	100.00	320.00	300.00	207.20	80.00
t_{23}	62.56	115.73	6.13	79.29	74.75	11.88	52.22	160.81	10.73	276.84	150.35	20.00	100.00	316.90	300.00	177.20	80.00
t_{24}	58.59	120.00	5.73	77.98	70.00	12.75	52.85	170.00	10.00	266.48	140.00	20.00	100.00	276.90	300.00	147.20	80.00

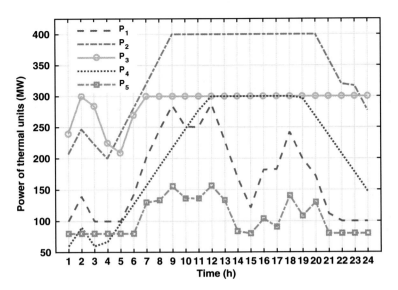

Fig. 3.8 Thermal unit schedules in multi-area economic dispatch

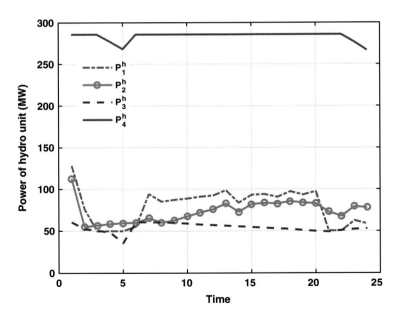

Fig. 3.9 Hydro unit schedules in multi-area economic dispatch

3.5 Multi-Area Mix Unit Dynamic Dispatch

The multi-area economic dispatch refers to a class of economic dispatch problems in which there are several interconnected areas. In each area, there are different power plant technologies and different demand pattern. The question is how to dispatch different power plants to supply the demand while minimizing the operating costs. A basic multi-area economic dispatch problem is shown in Fig. 3.10. There are ten thermal power plants which their connection area are given in Fig. 3.10. There exist two wind power plants, one in area 1 (250 MW) and one in area 3 (350 MW).

The multi-area dynamic economic dispatch problem is formulated in (3.16). The objective function is defined as total operating costs (3.16a). The operating limits of thermal units are given in (3.16b). The supply-demand balance of area a at time t is modeled in (3.16c). The amount of transferable power between area a and $a*$ is described in (3.16d). The actual wind dispatch ($pw_{w,t}$) depends on wind availability ($\xi_{w,t}$) and wind capacity ($\bar{pw}_w$) as expressed in (3.16e). $L^a_{e,t}$ is the electric load in area a and time t in (3.16c).

$$\min_{p_{i,t}, pw_{w,t}, \text{Tie}^{aa*}_t} \text{OF} = \sum_{i,t} a_i p^2_{i,t} + b_i p_{i,t} + c_i \tag{3.16a}$$

$$P^{\min}_i \leq p_{i,t} \leq P^{\max}_i \tag{3.16b}$$

$$\sum_w pw_{w,t} + \sum_i P_{i,t} = \sum_{a*} \text{Tie}^{aa*}_t + L^a_{e,t}, \forall i, w \in a, \forall t \tag{3.16c}$$

$$\left| \text{Tie}^{aa*}_t \right| \leq \bar{\text{Tie}}^{aa*} \tag{3.16d}$$

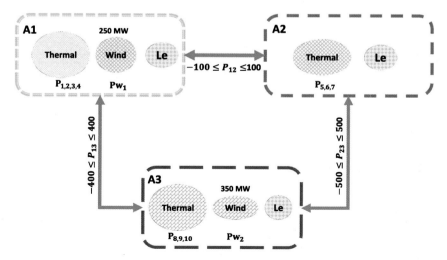

Fig. 3.10 Multi-area dynamic economic dispatch problem

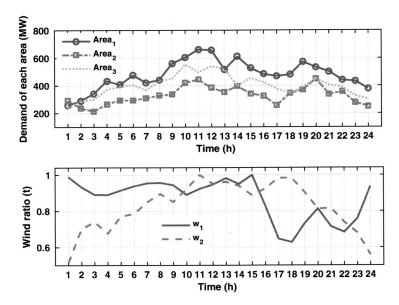

Fig. 3.11 Wind and demand pattern in each area

Table 3.12 Characteristics of thermal units and their connected area

Area	Unit	a_i ($/MW^2)	b_i ($/MW)	c_i ($)	$P_i^{\min}$ (MW)	$P_i^{\max}$ (MW)	RU$_i$ (MW)	RD$_i$ (MW)
A_1	P1	0.0056	17.87	601.75	20	150	40	40
	P2	0.0079	21.62	480.29	40	200	80	80
	P3	0.007	23.9	471.6	30	300	100	100
	P4	0.0043	21.6	958.2	30	350	120	120
A_2	P5	0.0095	22.54	692.4	10	70	30	30
	P6	0.009	19.58	455.6	20	80	40	40
	P7	0.0063	21.05	1313.6	40	450	150	150
A_3	P8	0.0048	23.23	639.4	50	130	50	50
	P9	0.0039	20.81	604.97	100	340	100	100
	P10	0.0021	16.51	502.7	40	130	60	60

$$0 \leq pw_{w,t} \leq \xi_{w,t}\bar{pw}_w \tag{3.16e}$$

The wind and demand patterns in each area are shown in Fig. 3.11. The techno-economic characteristics of thermal units and their connected area are given in Table 3.12. The GCode 3.8 solves the multi-area economic dynamic dispatch Example (3.16). Four sets are defined for time, area, thermal units, and wind turbines, then it is indicated which thermal unit belongs to each area. The table wind data describes the $\xi_{w,t}$ in (3.16e). The wind capacities and their connected areas are indicated next. Tie-line limits and thermal unit characteristics are defined next. It

should be mentioned that some data of demand and wind tables in GCode 3.8 are
omitted to save space.

GCode 3.8 Multi-area economic dynamic dispatch Example (3.16)

```
Sets  t  /t1*t24/, Area  /A1*A3/, i  /  g1*g10  /,w  /w1*w2/;  set
      AreaGen(area,i);
AreaGen(area,i)=no; AreaGen('A1',i)$(ord(i)<5)=yes;
AreaGen('A2',i)$(ord(i)>4 and ord(i)<8)=yes;
AreaGen('A3',i)$(ord(i)>7)=yes; alias(area,region);
Table  winddata(t,w)
      w1                      w2
t1    0.989432703003337   0.515610651974288
t2    0.932703003337041   0.701561065197429
t3    0.890989988876529   0.740128558310376
t4    0.889877641824249   0.676308539944904
t5    0.914905450500556   0.770431588613407
t6    0.937708565072303   0.785123966942149
t7    0.954393770856507   0.849403122130395
t8    0.956618464961068   0.895316804407714
t24   0.937152391546162   0.557392102846648  ;
parameter  Windcap(w)
/w1 250
w2   350/;
parameter  AreaWind(area,w);  AreaWind('A1','w1')=yes;  AreaWind('
   A3','w2')=yes;
table  Tielim(area,region)
      a1      a2      a3
a1    0       100     400
a2    100     0       500
a3    400     500     0;
table  gendata(i,*)  generator  cost  characteristics  and  limits
      a       b       c       Pmin    Pmax    RU      RD
g1    0.0056  17.87   601.75  20      150     40      40
g2    0.0079  21.62   480.29  40      200     80      80
g3    0.0070  23.9    471.6   30      300     100     100
g4    0.0043  21.6    958.2   30      350     120     120
g5    0.0095  22.54   692.4   10      70      30      30
g6    0.0090  19.58   455.6   20      80      40      40
g7    0.0063  21.05   1313.6  40      450     150     150
g8    0.0048  23.23   639.4   50      130     50      50
g9    0.0039  20.81   604.97  100     340     100     100
g10   0.0021  16.51   502.7   40      130     60      60;
table  demand(t,area)
      a1      a2      a3
t1    258     292     237
t2    291     237     289
t3    343     214     299
t4    435     267     371
t5    408     295     393
t6    477     293     406
t7    422     311     369
t10   606     422     555
```

```
t24   377   251   302  ;
variables  Tie(area,region,t),      OF,     p(i,t),     Pw(w,t);
tie.lo(area,region,t)=-Tielim(area,region);
tie.up(area,region,t)=+Tielim(area,region);
tie.fx(area,region,t)$(Tielim(area,region)=0)=0;
tie.fx(area,area,t)=0;Pw.lo(w,t)=0;
p.up(i,t)=gendata(i,"Pmax");p.lo(i,t)=gendata(i,"Pmin");
Pw.up(w,t)=winddata(t,w)*Windcap(w);
Equations  tieconst,balance,RampUp,RampDn,cost;
tieconst(area,region,t)  ..  Tie(area,region,t)=e=-Tie(region,
    area,t);
balance(area,t)..sum(i$AreaGen(area,i),p(i,t))+sum(w$AreaWind
    (area,w),Pw(w,t))=e=demand(t,area)+sum(region,Tie(area,
    region,t));
RampUp(i,t)..p(i,t)-p(i,t-1)=l=gendata(i,'RU');
RampDn(i,t)..p(i,t-1)-p(i,t)=l=gendata(i,'RD');
cost..  OF=e= sum((i,t),gendata(i,'a')*p(i,t)*p(i,t)+gendata(i,'
    b')*p(i,t)
+gendata(i,'c'));
Model edc /all/ ;
Solve edc min OF us QCP ;
```

The inter-area power transfer of different tie-lines is shown in Fig. 3.12.

Some other examples of multi-area dynamic dispatch can be found in [8]. The optimal solution of multi-area economic dispatch is described in Table 3.13.

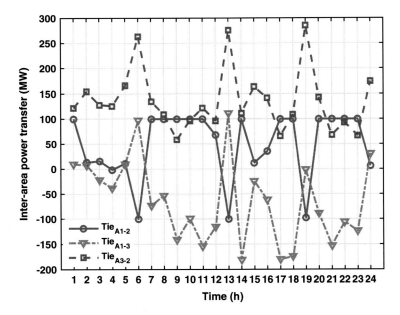

Fig. 3.12 Inter-area power transfer of different tie-lines

Table 3.13 Solution of multi-area economic dispatch

Time	P_1	P_2	P_3	P_4	P_5	P_6	P_7	P_8	P_9	P_{10}	Pw_1	Pw_2
t_1	20.0	40.0	30.0	30.0	10.0	20.0	40.0	50.0	100.0	40.0	247.4	159.6
t_2	20.0	40.0	30.0	30.0	10.0	20.0	40.0	50.0	100.0	40.0	191.5	245.5
t_3	20.0	40.0	30.0	30.0	10.0	20.0	40.0	50.0	100.0	40.0	217.0	259.0
t_4	60.0	40.0	30.0	42.7	10.0	60.0	72.8	50.0	148.3	100.0	222.5	236.7
t_5	100.0	40.0	30.0	30.0	10.0	67.6	40.0	50.0	100.0	130.0	228.7	269.7
t_6	140.0	40.0	30.0	30.0	10.0	66.1	53.5	50.0	117.2	130.0	234.4	274.8
t_7	110.0	40.0	30.0	30.0	10.0	26.1	40.0	50.0	100.0	130.0	238.6	297.3
t_8	150.0	40.0	30.0	30.0	10.0	66.1	43.1	50.0	100.3	130.0	239.2	313.4
t_9	150.0	40.0	30.0	65.8	10.0	80.0	88.6	50.0	173.9	130.0	236.0	297.8
t_{10}	150.0	71.2	30.0	133.2	10.8	80.0	134.5	50.0	248.1	130.0	222.5	322.7
t_{11}	150.0	69.1	30.0	129.3	10.0	80.0	131.9	50.0	243.9	130.0	230.8	350.0
t_{12}	150.0	67.5	30.0	126.4	10.0	80.0	129.9	50.0	240.7	130.0	236.7	332.8
t_{13}	150.0	40.0	30.0	64.2	10.0	80.0	87.5	50.0	172.1	130.0	245.6	336.7
t_{14}	150.0	40.0	30.0	74.2	10.0	80.0	94.3	50.0	183.1	130.0	237.5	328.9
t_{15}	150.0	40.0	30.0	46.5	10.0	80.0	75.4	50.0	152.6	130.0	250.0	310.5
t_{16}	150.0	40.0	30.0	30.0	10.0	80.0	58.3	50.0	124.9	130.0	207.3	323.5
t_{17}	128.2	40.0	30.0	30.0	10.0	40.0	40.0	50.0	100.0	130.0	161.7	344.1
t_{18}	150.0	40.0	30.0	30.0	10.0	80.0	45.3	50.0	104.0	130.0	157.0	343.7
t_{19}	150.0	40.0	30.0	71.7	10.0	80.0	92.6	50.0	180.3	130.0	183.0	316.4
t_{20}	150.0	56.2	30.0	105.6	10.0	80.0	115.7	50.0	217.7	130.0	203.3	284.4
t_{21}	150.0	40.0	30.0	53.2	10.0	80.0	80.0	50.0	159.9	130.0	178.5	285.4
t_{22}	150.0	40.0	30.0	44.4	10.0	80.0	73.9	50.0	150.2	130.0	170.9	258.6
t_{23}	120.2	40.0	30.0	30.0	10.0	60.0	40.0	50.0	100.0	130.0	189.8	238.0
t_{24}	80.2	40.0	30.0	30.0	10.0	20.0	40.0	50.0	100.0	100.4	234.3	195.1

3.6 Applications

Some optimization of economic dispatch models utilized in power system studies is given in this section:

- Economic dispatch with valve point loading [9]
- Non-smooth cost functions in economic dispatch analysis [10]
- Multi fuel handling [11]
- Transmission loss impacts on ED problem [12]
- Reserve constrained economic dispatch with prohibited operating zones [13]
- Environmentally constrained economic dispatch [14]
- Security constrained economic dispatch problem [15]
- Frequency-constrained stochastic ED [16]
- Decentralized economic dispatch [17]

Nomenclature

Indices and Sets

i	Index of thermal generating units.
j	Index of chp units.
k	Index of heat only units.
Ω_{th}	Set of all thermal generating units.
Ω_h	Set of all heat only units.
Ω_{chp}	Set of all chp units.

Parameters

P_i^{th}	Generated electric power by thermal unit i.
p_j^{chp}	Generated electric power by chp unit j.
q_j^{chp}	Generated heat power by chp unit j.
F^{th}	Costs of thermal units.
F^h	Costs of heat only units.
F^{chp}	Costs of chp units.
q_k^h	Generated heat power by heat only unit k.
$Q_k^{h,\min/\max}$	Minimum and maximum limits of heat generation of heat only unit k.
$P_j^{\text{chp},\min/\max}$	Minimum and maximum limits of electricity generation of chp unit j.
$Q_j^{\text{chp},\min/\max}$	Minimum and maximum limits of heat generation of chp unit j.
L_e	Electric power demand
L_h	Heat power demand
$\{d,e,f\}_i^{\text{th}}$	Emission coefficients of thermal unit i.
$\{a,b,c\}_i^{\text{th}}$	Fuel cost coefficients of thermal unit i.
$\{a,b,c\}_i^h$	Fuel cost coefficients of heat unit k.
$\{a,b,c,d,e,f\}_j^{\text{chp}}$	Fuel cost coefficients of chp unit j.
$P_i^{\text{th},\max/\min}$	Maximum/minimum limits of power generation of thermal unit i.
Elimit	Maximum emission limit (kg).
ϵ	Parameter used for multi-objective optimization.
C_e	Emission price ($/kg).

Variables

P_i	Power generated by thermal unit i (MW).
OF	Total operating costs ($).
TC	Total fuel cost ($).
EM	Total emissions (kg).

References

1. A. Rabiee, A. Soroudi, B. Mohammadi-Ivatloo, M. Parniani, Corrective voltage control scheme considering demand response and stochastic wind power. IEEE Trans. Power Syst. **29**(6), 2965–2973 (2014)
2. T. Guo, M.I. Henwood, M. van Ooijen, An algorithm for combined heat and power economic dispatch. IEEE Trans. Power Syst. **11**(4), 1778–1784 (1996)
3. A.J. Conejo, J.M. Arroyo, J. Contreras, F.A. Villamor, Self-scheduling of a hydro producer in a pool-based electricity market. IEEE Trans. Power Syst. **17**(4), 1265–1272 (2002)
4. J. Zhang, J. Wang, C. Yue, Small population-based particle swarm optimization for short-term hydrothermal scheduling. IEEE Trans. Power Syst. **27**(1), 142–152 (2012)
5. M.E. Khodayar, M. Shahidehpour, L. Wu, Enhancing the dispatchability of variable wind generation by coordination with pumped-storage hydro units in stochastic power systems. IEEE Trans. Power Syst. **28**(3), 2808–2818 (2013)
6. A.A. Snchez de la Nieta, J. Contreras, J.I. Muoz, Optimal coordinated wind-hydro bidding strategies in day-ahead markets. IEEE Trans. Power Syst. **28**(2), 798–809 (2013)
7. A. Soroudi, Robust optimization based self scheduling of hydro-thermal Genco in smart grids. Energy **61**, 262–271 (2013)
8. A. Soroudi, A. Rabiee, Optimal multi-area generation schedule considering renewable resources mix: a real-time approach. IET Gener. Transm. Distrib. **7**(9), 1011–1026 (2013)
9. D.C. Walters, G.B. Sheble, Genetic algorithm solution of economic dispatch with valve point loading. IEEE Trans. Power Syst. **8**(3), 1325–1332 (1993)
10. J.-B. Park, K.-S. Lee, J.-R. Shin, K.Y. Lee, A particle swarm optimization for economic dispatch with nonsmooth cost functions. IEEE Trans. Power Syst. **20**(1), 34–42 (2005)
11. C.-L. Chiang, Improved genetic algorithm for power economic dispatch of units with valve-point effects and multiple fuels. IEEE Trans. Power Syst. **20**(4), 1690–1699 (2005)
12. Z.X. Liang, J.D. Glover, A zoom feature for a dynamic programming solution to economic dispatch including transmission losses. IEEE Trans. Power Syst. **7**(2), 544–550 (1992)
13. F.N. Lee, A.M. Breipohl, Reserve constrained economic dispatch with prohibited operating zones. IEEE Trans. Power Syst. **8**(1), 246–254 (1993)
14. K.P. Wang, J. Yuryevich, Evolutionary-programming-based algorithm for environmentally-constrained economic dispatch. IEEE Trans. Power Syst. **13**(2), 301–306 (1998)
15. R.A. Jabr, A.H. Coonick, B.J. Cory, A homogeneous linear programming algorithm for the security constrained economic dispatch problem. IEEE Trans. Power Syst. **15**(3), 930–936 (2000)
16. Y.Y. Lee, R. Baldick, A frequency-constrained stochastic economic dispatch model. IEEE Trans. Power Syst. **28**(3), 2301–2312 (2013)
17. W.T. Elsayed, E.F. El-Saadany, A fully decentralized approach for solving the economic dispatch problem. IEEE Trans. Power Syst. **30**(4), 2179–2189 (2015)

Chapter 4
Dynamic Economic Dispatch

This chapter provides a solution for dynamic economic dispatch (DED) problem in GAMS. DED refers to dispatch a set of units over a given operating horizon (usually 24 h). The on/off status of units is assumed to be known. Two paradigms of DED are discussed and modeled, namely cost-based DED and price-based DED. Finally, the linearized version of DED is provided.

4.1 Cost-Based DED

The objective function of cost-based DED is minimizing the total operating costs while satisfying the hourly demands and other technical constraints. The production costs of g-th thermal unit is defined as:

$$C(P_{g,t}) = a_g P_{g,t}^2 + b_g P_{g,t} + c_g \tag{4.1}$$

where a_g, b_g, and c_g are the fuel cost coefficients of the g-th unit. The total fuel cost (TC) is calculated as follows:

$$\text{TC} = \sum_{g,t} C(P_{g,t}) \tag{4.2}$$

The total emission (EM) is calculated as follows:

$$\text{EM} = \sum_{g,t} d_g P_{g,t}^2 + e_g P_{g,t} + f_g \tag{4.3}$$

The operating limits are defined as follows:

$$P_g^{\min} \leq P_{g,t} \leq P_g^{\max} \tag{4.4}$$

© Springer International Publishing AG 2017
A. Soroudi, *Power System Optimization Modeling in GAMS*,
DOI 10.1007/978-3-319-62350-4_4

Table 4.1 The dynamic economic dispatch data for four units example

g	a_g ($/MW^2)	b_g ($/MW)	c_g ($)	d_g (kg/MW^2)	e_g (kg/MW)	f_g (kg)	P_g^{min} (MW)	P_g^{max} (MW)	RU_g (MW)	RD_g (MW)
$g1$	0.12	14.8	89	1.2	-5	3	28	200	40	40
$g2$	0.17	16.57	83	2.3	-4.24	6.09	20	290	30	30
$g3$	0.15	15.55	100	1.1	-2.15	5.69	30	190	30	30
$g4$	0.19	16.21	70	1.1	-3.99	6.2	20	260	50	50

where $P_g^{max/min}$ are the maximum/minimum power outputs of g-th thermal unit. The overall thermal unit cost-based dynamic economic dispatch is formulated as follows:

$$\min_{P_{g,t}} TC = \sum_{g,t} a_g P_{g,t}^2 + b_g P_{g,t} + c_g \qquad (4.5a)$$

$$EM = \sum_{g,t} d_g P_{g,t}^2 + e_g P_{g,t} + f_g \qquad (4.5b)$$

$$P_g^{min} \leq P_{g,t} \leq P_g^{max} \qquad (4.5c)$$

$$P_{g,t} - P_{g,t-1} \leq RU_g \qquad (4.5d)$$

$$P_{g,t-1} - P_{g,t} \leq RD_g \qquad (4.5e)$$

$$\sum_g P_{g,t} \geq L_t \qquad (4.5f)$$

The economic dispatch data for four units example is given in Table 4.1. This table has 11 columns. The first column is showing the generating unit index. The next three columns indicate the cost coefficients for these thermal units (a_g, b_g, c_g). The next three columns indicate the emission coefficients for these thermal units (d_g, e_g, f_g). The next two columns give the minimum and maximum generating limits of each unit if they are on. The last two columns indicate the ramp up/down rates of thermal units.

The GAMS code for solving the example (4.5) is given in GCode 4.1.

GCode 4.1 The cost-based dynamic economic dispatch Example (4.5)

```
Sets   t          hours              /t1*t24/
       i          thermal  units     /g1*g4/;
table  gendata(i,*)  generator  cost  characteristics  and  limits
       a       b      c    d     e      f     Pmin  Pmax  RU0   RD0
g1     0.12   14.80  89   1.2  -5       3     28    200   40    40
g2     0.17   16.57  83   2.3  -4.24  6.09    20    290   30    30
g3     0.15   15.55  100  1.1  -2.15  5.69    30    190   30    30
g4     0.19   16.21  70   1.1  -3.99  6.2     20    260   50    50;
Parameter  demand(t)
/t1          510
t2           530
t3           516
```

```
t4          510
t5          515
t6          544
t7          646
t8          686
t9          741
t10         734
t11         748
t12         760
t13         754
t14         700
t15         686
t16         720
t17         714
t18         761
t19         727
t20         714
t21         618
t22         584
t23         578
t24         544/;
Variables         OBJ             Objective (revenue)
                  costThermal     Cost of thermal units
                  p(i,t)          Power generated by thermal power
                                  plant
                  EM              Emission calculation         ;
p.up(i,t) = gendata(i,"Pmax") ;
p.lo(i,t) = gendata(i,"Pmin");

Equations Genconst3, Genconst4, costThermalcalc, balance, EMcalc;

costThermalcalc .. costThermal =e=sum((t,i), gendata(i,'a')*power(
    p(i,t),2)
+gendata(i,'b')*p(i,t) +gendata(i,'c'));
Genconst3(i,t) .. p(i,t+1)−p(i,t)=l=gendata(i,'RU0');
Genconst4(i,t) .. p(i,t−1)−p(i,t)=l=gendata(i,'RD0');
balance(t) .. sum(i,p(i,t))=g=demand(t);
EMcalc     .. EM=e=sum((t,i), gendata(i,'d')*power(p(i,t),2)
+gendata(i,'e')*p(i,t) +gendata(i,'f'));

Model DEDcostbased / all /;

Solve DEDcostbased us QCP min costThermal;

execute_unload "DEDcostbased.gdx" P.l
execute 'gdxxrw.exe DEDcostbased.gdx var=P  rng=Pthermal!al'
```

The thermal unit power schedules obtained by GCode 4.1 are shown in Fig. 4.1.

The numerical values of dynamic economic dispatch solution for four units example are given in Table 4.2. The total operating costs are TC = 6.4796×10^5. The total emissions are 3.5929×10^3 tons.

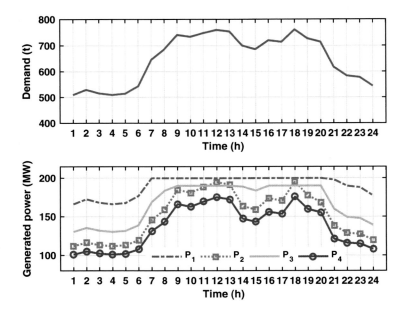

Fig. 4.1 The hourly thermal unit power schedules in cost-based DED

4.1.1   Ramp Rate Sensitivity Analysis

In this section, the impact of ramp rates on optimal solution will be investigated.
For this reason, the ramp rates are gradually (in 2% steps) reduced to 60% of their
original values. The GAMS code for solving the ramp sensitivity example is given
in GCode 4.2.

GCode 4.2 The ramp rate sensitivity analysis of cost-based dynamic economic dispatch

```
Sets        t     /t1*t24/,  i   /p1*p4/;
Table  gendata(i,*)  generator  cost  characteristics  and  limits
           a      b      c     d     e      f    Pmin  Pmax  RU0   RD0
p1       0.12  14.80  89   1.2  −5     3    28    200   40    40
p2       0.17  16.57  83   2.3  −4.24  6.09 20    290   30    30
p3       0.15  15.55  100  1.1  −2.15  5.69 30    190   30    30
p4       0.19  16.21  70   1.1  −3.99  6.2  20    260   50    50;
parameter  demand(t)
/t1          510
t2           530
t3           516
t4           510
t5           515
t6           544
t7           646
t8           686
```

```
t9              741
t10             734
t11             748
t12             760
t13             754
t14             700
t15             686
t16             720
t17             714
t18             761
t19             727
t20             714
t21             618
t22             584
t23             578
t24             544
/;
Variables          OBJ              Objective (revenue)
                   costThermal      Cost of thermal units
                   p(i,t)           Power generated by the thermal
                                    power plant
                   EM               Emission calculation;
p.up(i,t) = gendata(i,"Pmax") ;
p.lo(i,t) = gendata(i,"Pmin");
Equations Genconst3,Genconst4,costThermalcalc,balance,EMcalc;
costThermalcalc.. costThermal =e=sum((t,i), gendata(i,'a')*power
(p(i,t),2)+gendata(i,'b')*p(i,t) +gendata(i,'c'));
Genconst3(i,t) .. p(i,t+1)-p(i,t)=l=gendata(i,'RU');
Genconst4(i,t) .. p(i,t-1)-p(i,t)=l=gendata(i,'RD');
balance(t) .. sum(i,p(i,t))=g=demand(t);
EMcalc .. EM=e=sum((t,i), gendata(i,'d')*power(p(i,t),2)+gendata
(i,'e')*p(i,t)+gendata(i,'f'));
Model DEDcostbased /all/;
Scalar Rscale /1/;
set counter /c1*c21/;
parameter report1(counter,*);
loop(counter,
Rscale=1-(ord(counter)-1)*0.02;
gendata(i,'RU')=gendata(i,'RU0')*RScale;
gendata(i,'RD')=gendata(i,'RD0')*RScale;
Solve DEDcostbased us qcp min costThermal;
report1(counter,'Scale')=Rscale;
report1(counter,'TC')=costThermal.l;
report1(counter,'EM')=EM.l;
);
display report1;
execute_unload "DEDcostbased.gdx" report1
execute 'gdxxrw.exe DEDcostbased.gdx par=report1 rng=Pthermal!a1'
```

The sensitivity analysis of ramp rates on hourly thermal unit power schedules in DED is shown in Fig. 4.2.

Table 4.2 The dynamic economic dispatch solution for four units expressed in (MW)

Time	p_1	p_2	p_3	p_4	Load
t_1	166.20	112.10	130.50	101.30	510.00
t_2	172.60	116.60	135.60	105.30	530.00
t_3	168.10	113.50	132.00	102.50	516.00
t_4	166.20	112.10	130.50	101.30	510.00
t_5	167.80	113.20	131.70	102.30	515.00
t_6	177.00	119.80	139.10	108.10	544.00
t_7	200.00	145.90	168.70	131.50	646.00
t_8	200.00	159.10	183.70	143.30	686.00
t_9	200.00	184.80	190.00	166.30	741.00
t_{10}	200.00	181.10	190.00	162.90	734.00
t_{11}	200.00	188.40	190.00	169.60	748.00
t_{12}	200.00	194.80	190.00	175.20	760.00
t_{13}	200.00	191.60	190.00	172.40	754.00
t_{14}	200.00	163.70	188.90	147.40	700.00
t_{15}	200.00	159.10	183.70	143.30	686.00
t_{16}	200.00	173.70	190.00	156.30	720.00
t_{17}	200.00	170.50	190.00	153.50	714.00
t_{18}	200.00	195.30	190.00	175.70	761.00
t_{19}	200.00	177.40	190.00	159.60	727.00
t_{20}	200.00	168.60	190.00	155.40	714.00
t_{21}	198.00	138.60	160.00	121.40	618.00
t_{22}	189.80	128.80	149.30	116.20	584.00
t_{23}	187.90	127.40	147.80	114.90	578.00
t_{24}	177.00	119.80	139.10	108.10	544.00

The numerical results for ramp rate sensitivity analysis in dynamic economic dispatch of 4 units are given in Table 4.3. As it is seen in Fig. 4.2, the total costs will increase by decreasing the ramp rates. The final total costs (40% reduction in ramp rates) will be $649,226.53.

4.1.2   Multi-Objective Cost-Emission Minimization

In this section, the decision maker needs to minimize the emissions and total operating costs simultaneously.

$$\min_{P_{g,t}} TC = \sum_{g,t} a_g P_{g,t}^2 + b_g P_{g,t} + c_g \tag{4.6a}$$

$$\min_{P_{g,t}} EM = \sum_{g,t} d_g P_{g,t}^2 + e_g P_{g,t} + f_g \tag{4.6b}$$

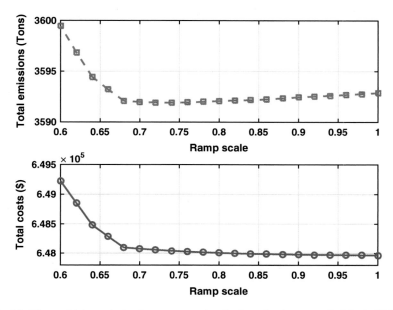

Fig. 4.2 The sensitivity analysis of ramp rates on hourly thermal unit power schedules in DED

Table 4.3 The numerical results for ramp rate sensitivity analysis in dynamic economic dispatch of 4 units

Counter	Scale	TC ($)	EM (kg)
c_1	1.00	647,964.46	3,592,886.80
c_2	0.98	647,966.18	3,592,796.25
c_3	0.96	647,968.28	3,592,707.78
c_4	0.94	647,970.78	3,592,622.76
c_5	0.92	647,973.69	3,592,545.64
c_6	0.90	647,977.11	3,592,469.20
c_7	0.88	647,981.17	3,592,362.06
c_8	0.86	647,985.94	3,592,261.31
c_9	0.84	647,991.45	3,592,183.41
c_{10}	0.82	647,997.77	3,592,150.64
c_{11}	0.80	648,005.10	3,592,086.32
c_{12}	0.78	648,014.58	3,592,007.06
c_{13}	0.76	648,026.42	3,591,950.09
c_{14}	0.74	648,040.63	3,591,919.82
c_{15}	0.72	648,057.21	3,591,935.58
c_{16}	0.70	648,076.19	3,591,985.60
c_{17}	0.68	648,098.11	3,592,096.38
c_{18}	0.66	648,289.75	3,593,218.29
c_{19}	0.64	648,484.76	3,594,442.68
c_{20}	0.62	648,853.35	3,596,838.57
c_{21}	0.60	649,226.53	3,599,445.17

$$P_g^{\min} \leq P_{g,t} \leq P_g^{\max} \tag{4.6c}$$

$$P_{g,t} - P_{g,t-1} \leq \mathrm{RU}_g \tag{4.6d}$$

$$P_{g,t-1} - P_{g,t} \leq \mathrm{RD}_g \tag{4.6e}$$

$$\sum_g P_{g,t} \geq L_t \tag{4.6f}$$

The procedure is as follows:

1. Find the maximum of each objective function and save them.
2. Add one of the objective functions (here EM) to the constraints as follows:

$$\mathrm{EM} \leq \epsilon \tag{4.7}$$

3. The ϵ value will be varied from $\mathrm{EM}^{\max}$ to $\mathrm{EM}^{\min}$ and then TC is minimized.

The Pareto optimal front for emission-cost minimization DED is shown in Fig. 4.3. The numerical results for Pareto optimal solution are provided in Table 4.4. The question is which solution of the Pareto optimal front should be chosen as the best solution? Which one is the "most preferred" solution among those located on Pareto optimal front of Fig. 4.3?. A fuzzy satisfying method [1] is used in this section to find the "the best" solution. The principles of this method are as follows: for each solution in the Pareto optimal front, X_c, a membership function is defined as $\mu^{f_k(X_c)}$. This value, which varies between 0 to 1, shows the success degree of X_c in minimizing the objective function k. A linear membership function is used for both objective functions, as follows:

$$\mu^{f_k(X_c)} = \begin{cases} 0 & \text{Otherwise} \\ \frac{f_k^{\max} - f_k(X_c)}{f_k^{\max} - f_k^{\min}} & f_k^{\min} \leq f_k(X_c) \leq f_k^{\max} \end{cases} \tag{4.8}$$

A conservative decision maker tries to maximize minimum satisfaction among all objectives or minimize the maximum dissatisfaction. The final solution can then be found as:

$$\max_{c=1:11} \left(\min_{k=1:2} \mu^{f_k(X_c)} \right) \tag{4.9}$$

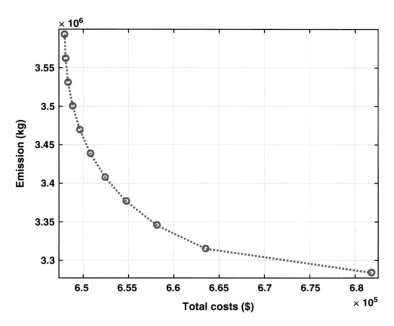

Fig. 4.3 The Pareto optimal front for emission-cost minimization DED

GCode 4.3 The multi-objective cost-emission minimization of dynamic economic dispatch

```
sets  t   /t1*t24/,   i  /g1*g4/,  counter  /c1*c11/;  Scalar  lim  /inf/;
table  gendata(i,*)  generator  cost  characteristics  and  limits
        a      b       c     d      e      f     Pmin  Pmax  RU0    RD0
g1    0.12  14.80   89   1.2   -5     3     28    200   40     40
g2    0.17  16.57   83   2.3   -4.24  6.09  20    290   30     30
g3    0.15  15.55  100   1.1   -2.15  5.69  30    190   30     30
g4    0.19  16.21   70   1.1   -3.99  6.2   20    260   50     50;
parameter  demand(t)
/t1          510
t2           530
t3           516
t4           510
t5           515
t6           544
t7           646
t8           686
t9           741
t10          734
t11          748
t12          760
t13          754
t14          700
t15          686
```

```
t16        720
t17        714
t18        761
t19        727
t20        714
t21        618
t22        584
t23        578
t24        544/;
Variables            OBJ, costThermal , p( i , t ) ,EM  ;
p.up( i , t ) = gendata( i ,"Pmax"); p.lo( i , t ) = gendata( i ,"Pmin");
Equations  Genconst3 , Genconst4 , costThermalcalc , balance , EMcalc ,
    EMlim ;
costThermalcalc .. costThermal =e=sum(( t , i ) , gendata( i , 'a')*power
( p( i , t ) ,2)+gendata( i , 'b')*p( i , t ) +gendata( i , 'c'));
Genconst3 ( i , t )  .. p( i , t+1)−p( i , t )=l=gendata( i , 'RU0') ;
Genconst4 ( i , t )  .. p( i , t −1)−p( i , t )=l=gendata( i , 'RD0') ;
balance( t )  .. sum( i , p( i , t ) )=e=demand( t );
EMcalc .. EM=e=sum(( t , i ) , gendata( i , 'd')*power(p( i , t ) ,2)+gendata
( i , 'e')*p( i , t )+gendata( i , 'f'));
EMlim    .. EM =l=lim ;
Model DEDcostbased  / all /;
parameter  report1 ( counter ,*) , rep (*) , report2 ( counter , i , t ) ;
Solve DEDcostbased us qcp min costThermal ;
rep ( 'TCmin')=costThermal . l ;
rep ( 'EMmax')=EM. l ;
Solve DEDcostbased us qcp min EM;
rep ( 'TCmax')=costThermal . l ;
rep ( 'EMmin')=EM. l ;
loop( counter ,
lim=rep ( 'EMmax')+( rep ( 'EMmin')−rep ( 'EMmax') )*( ord ( counter )−1)/
    ( card ( counter )−1);
Solve DEDcostbased us qcp min costThermal ;
report1 ( counter , 'Epsilon')=lim ;
report1 ( counter , 'TC')=costThermal . l ;
report1 ( counter , 'EM')=EM. l ;
report2 ( counter , i , t )=P. l ( i , t );
);
execute_unload "DEDcostbased. gdx" report1
execute 'gdxxrw. exe DEDcostbased. gdx par=report1   rng=report!a1'
execute_unload "DEDcostbased. gdx" report2
execute 'gdxxrw. exe DEDcostbased. gdx par=report2   rng=Pthermal!a1'
```

Considering the (4.9) and the values of Table 4.4, the best solution is $c9$. The costs and emissions of this solution are \$658,087.973 and 3.3459×10^3 tones, respectively. The best compromised solution of Pareto optimal front for emission-cost minimization DED is shown in Fig. 4.4.

Table 4.4 The numerical results for Pareto optimal solutions (cost-based DED)

Solution	ϵ	TC ($)	EM (kg)	μ_{TC}	μ_{EM}	$\min(\mu_{TC}, \mu_{EM})$
$c1$	3,592,886.799	647,964.460	3,592,886.799	1.000	0.000	0.000
$c2$	3,562,019.341	648,050.068	3,562,019.341	0.997	0.100	0.100
$c3$	3,531,151.883	648,328.132	3,531,151.883	0.989	0.200	0.200
$c4$	3,500,284.425	648,836.461	3,500,284.425	0.974	0.300	0.300
$c5$	3,469,416.967	649,627.557	3,469,416.967	0.951	0.400	0.400
$c6$	3,438,549.510	650,778.132	3,438,549.510	0.917	0.500	0.500
$c7$	3,407,682.052	652,408.206	3,407,682.052	0.869	0.600	0.600
$c8$	3,376,814.594	654,716.176	3,376,814.594	0.800	0.700	0.700
$c9$	3,345,947.136	658,087.973	3,345,947.136	0.701	0.800	0.701
$c10$	3,315,079.678	663,531.693	3,315,079.678	0.540	0.900	0.540
$c11$	3,284,212.220	681,788.377	3,284,212.220	0.000	1.000	0.000

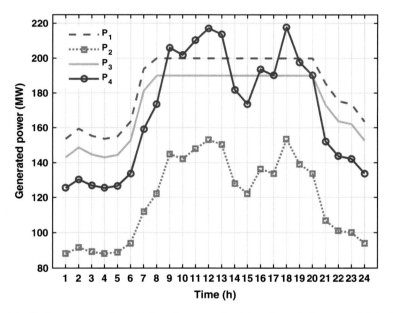

Fig. 4.4 The best compromised solution of Pareto optimal front for emission-cost minimization DED

4.1.3 Wind Integrated DED

In this section, the role of wind power generation in DED problem is investigated. The integration of wind power generation in DED problem is formulated as follows:

$$\min_{DV} TC = \sum_{g,t} a_g P_{g,t}^2 + b_g P_{g,t} + c_g + \sum_{t} VWC \times P_t^{wc} \qquad (4.10a)$$

$$DV = \{P_{g,t}, P_t^w, P_t^{wc}\} \tag{4.10b}$$

$$P_g^{\min} \leq P_{g,t} \leq P_g^{\max} \tag{4.10c}$$

$$P_{g,t} - P_{g,t-1} \leq RU_g \tag{4.10d}$$

$$P_{g,t-1} - P_{g,t} \leq RD_g \tag{4.10e}$$

$$P_t^w + \sum_g P_{g,t} \geq L_t \tag{4.10f}$$

$$P_t^w + P_t^{wc} \leq \Lambda_t^w \tag{4.10g}$$

The GAMS code for solving the example (4.10) is given in GCode 4.4.

GCode 4.4 The cost-based wind-DED (4.10)

```
Sets       t          hours          /t1*t24/,
           g          thermal units  /p1*p4/;

table  gendata(g,*) generator cost characteristics and limits
      a     b     c    d    e     f    Pmin Pmax  RU0   RD0
p1    0.12  14.80 89   1.2  -5    3    28   200   40    40
p2    0.17  16.57 83   2.3  -4.24 6.09 20   290   30    30
p3    0.15  15.55 100  1.1  -2.15 5.69 30   190   30    30
p4    0.19  16.21 70   1.1  -3.99 6.2  20   260   50    50;
table  data(t,*)
      lambda load  wind
t1    32.71  510   44.1
t2    34.72  530   48.5
t3    32.71  516   65.7
t4    32.74  510   144.9
t5    32.96  515   202.3
t6    34.93  544   317.3
t7    44.9   646   364.4
t8    52     686   317.3
t9    53.03  741   271
t10   47.26  734   306.9
t11   44.07  748   424.1
t12   38.63  760   398
t13   39.91  754   487.6
t14   39.45  700   521.9
t15   41.14  686   541.3
t16   39.23  720   560
t17   52.12  714   486.8
t18   40.85  761   372.6
t19   41.2   727   367.4
t20   41.15  714   314.3
t21   45.76  618   316.6
t22   45.59  584   311.4
t23   45.56  578   405.4
t24   34.72  544   470.4;
Variables            OBJ            Objective (revenue)
```

```
                        cost      Cost of thermal units
                        p(g,t)              Power generated by thermal
                                            power plant
                        Pw(t), PWC(t)    winf power wind curtailmet       ;
Pw.up(t)=data(t,'wind'); Pw.lo(t)=0;
Pwc.up(t)=data(t,'wind'); Pwc.lo(t)=0;
p.up(g,t) = gendata(g,"Pmax") ;
p.lo(g,t) = gendata(g,"Pmin");
scalar VWC /50/;
Equations Genconst3,Genconst4,costThermalcalc,balance, wind;
costThermalcalc..  cost=e=sum(t,VWC*pwc(t))+sum((t,g),gendata(g,
'a')*power(p(g,t),2) +gendata(g,'b')*p(g,t)      +gendata(g,'c')));
Genconst3(g,t) ..  p(g,t+1)-p(g,t)=l=gendata(g,'RU0');
Genconst4(g,t) ..  p(g,t-1)-p(g,t)=l=gendata(g,'RD0');
balance(t) ..        sum(g,p(g,t))+Pw(t)=g=data(t,'load');
wind(t) ..        Pw(t)+PWC(t)=e=data(t,'wind');
Model DEDwindcostbased /all/;
Solve DEDwindcostbased us qcp min cost;
parameter rep(t,*);
rep(t,'Pth')=sum(g,p.l(g,t));
rep(t,'PW')=PW.l(t);
rep(t,'Pwc')=PWc.l(t);
rep(t,'Load')=data(t,'load');
execute_unload "DEDwindcostbased.gdx" P.l
execute 'gdxxrw.exe DEDwindcostbased.gdx var=P  rng=Pthermal!a1'
execute_unload "DEDwindcostbased.gdx" rep
execute 'gdxxrw.exe DEDwindcostbased.gdx par=reP  rng=rep!a1'
```

The total operating costs with wind integration are TC = 2.2692×10^5 as obtained in GCode 4.4. The hourly schedules of thermal units in the wind-DED problem are shown in Fig. 4.5. The hourly variation of total wind and thermal power generation as well as the demand in wind-DED problem is depicted in Fig. 4.6.

4.2 Price-Based DED

The objective function of price-based DED (PBDED) is maximizing the total benefits of generating company by selling the energy or other products. In this case, there is no hourly demands-supply balance constraint; however, other technical constraints should still be satisfied. The net benefit of generating company is equal to income minus operating costs.

4.2.1 Price-Based DED Just Energy Market

This case is devoted to the case of price-based DED without arbitrage. This means that the only sold commodity is energy. The problem is formulated in (4.11) as follows:

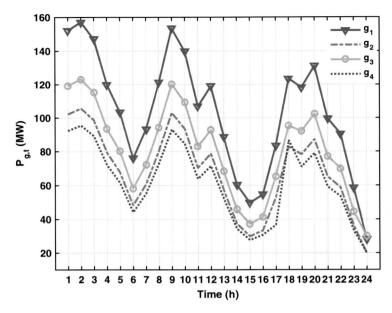

Fig. 4.5 The hourly schedules of thermal units in wind-DED problem

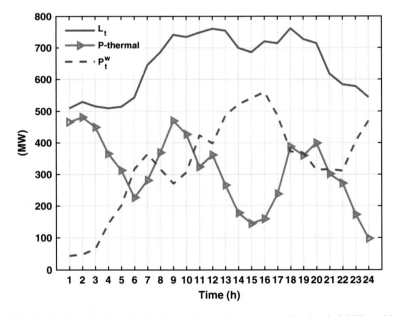

Fig. 4.6 The hourly variation of total wind and thermal power generation in wind-DED problem

$$\max_{P_{g,t}} \text{OF} = \sum_{g,t} \lambda_t^e P_{g,t} - \text{TC} \tag{4.11a}$$

$$\text{TC} = \sum_{g,t} a_g P_{g,t}^2 + b_g P_{g,t} + c_g \tag{4.11b}$$

$$\text{EM} = \sum_{g,t} d_g P_{g,t}^2 + e_g P_{g,t} + f_g \tag{4.11c}$$

$$P_g^{\min} \le P_{g,t} \le P_g^{\max} \tag{4.11d}$$

$$P_{g,t} - P_{g,t-1} \le \text{RU}_g \tag{4.11e}$$

$$P_{g,t-1} - P_{g,t} \le \text{RD}_g \tag{4.11f}$$

$$\sum_{g,t} P_{g,t} \le L_t \tag{4.11g}$$

The objective function in (4.11a) models the benefits of Genco which is income minus the total operating costs. λ_t^e is the electricity price at time t in (4.11a). In this formulation, we just calculate the total emission in (4.11c). Another difference of price-based DED with cost-based DED is in Eq. (4.11g). As it can be observed in this equation, the total power should be less than demand at any time. This means that the maximum power that Genco can sell is limited to the demand at time t.

GCode 4.5 The code for price-based dynamic economic dispatch

```
Sets       t          hours            /t1*t24/
           i          thermal units    /g1*g4/;
scalar lim /inf/;
table gendata(i,*) generator cost characteristics and limits
      a     b     c    d     e     f    Pmin Pmax  RU0   RD0
g1    0.12  14.80 89   1.2  −5     3    28   200   40    40
g2    0.17  16.57 83   2.3  −4.24  6.09 20   290   30    30
g3    0.15  15.55 100  1.1  −2.15  5.69 30   190   30    30
g4    0.19  16.21 70   1.1  −3.99  6.2  20   260   50    50;
table data(t,*)
           lambda        load
t1         32.71         510
t2         34.72         530
t3         32.71         516
t4         32.74         510
t5         32.96         515
t6         34.93         544
t7         44.9          646
t8         52            686
t9         53.03         741
t10        47.26         734
t11        44.07         748
t12        38.63         760
t13        39.91         754
t14        39.45         700
```

```
t15        41.14           686
t16        39.23           720
t17        52.12           714
t18        40.85           761
t19        41.2            727
t20        41.15           714
t21        45.76           618
t22        45.59           584
t23        45.56           578
t24        34.72           544;
Variables          OF              Objective (revenue)
                   costThermal     Cost of thermal units
                   p(i,t)          Power generated by thermal power
                         plant
                   EM              Emission calculation       ;
p.up(i,t) = gendata(i,"Pmax") ;
p.lo(i,t) = gendata(i,"Pmin");
equations
Genconst3, Genconst4, costThermalcalc, balance, EMcalc, EMlim,
     benefitcalc;
costThermalcalc.. costThermal =e=sum((t,i), gendata(i,'a')*power(
     p(i,t),2)
+gendata(i,'b')*p(i,t) +gendata(i,'c'));
Genconst3(i,t) .. p(i,t+1)-p(i,t)=l=gendata(i,'RU0');
Genconst4(i,t) .. p(i,t-1)-p(i,t)=l=gendata(i,'RD0');
balance(t) .. sum(i,p(i,t))=l=data(t,'load');
EMcalc .. EM=e=sum((t,i),gendata(i,'d')*power(p(i,t),2)+gendata(i
     ,'e')*p(i,t)
+gendata(i,'f'));
EMlim .. EM =l=lim;
benefitcalc .. OF=e=sum((i,t),1*data(t,'lambda')*p(i,t))-
     costThermal;
Model DEDPB /all/;
Solve DEDPB us qcp max of;
execute_unload "DEDPB.gdx" P.l
execute 'gdxxrw.exe DEDPB.gdx var=P rng=report!a1'
```

The optimal solution of price-based DED is depicted in Fig. 4.7.

4.2.2 Price-Based DED Energy and Reserve Market

This case is devoted to the case of price-based DED without arbitrage. This means that the only sold commodities are energy and spinning reserve. The problem is formulated in (4.12) as follows:

$$\max_{P_{g,t}} \text{OF} = \sum_{g,t} \left\{ \lambda_t^e P_{g,t} + \lambda_t^r \text{SR}_{g,t} \right\} - \text{TC} \tag{4.12a}$$

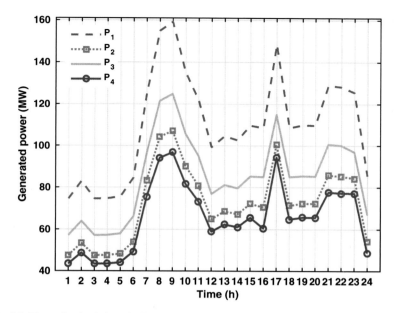

Fig. 4.7 The optimal solution of price-based DED

$$\mathrm{TC} = \sum_{g,t} a_g P_{g,t}^2 + b_g P_{g,t} + c_g \tag{4.12b}$$

$$\mathrm{EM} = \sum_{g,t} d_g P_{g,t}^2 + e_g P_{g,t} + f_g \tag{4.12c}$$

$$P_g^{\min} \le P_{g,t} \le P_g^{\max} \tag{4.12d}$$

$$P_{g,t} - P_{g,t-1} \le \mathrm{RU}_g \tag{4.12e}$$

$$P_{g,t-1} - P_{g,t} \le \mathrm{RD}_g \tag{4.12f}$$

$$0 \le \mathrm{SR}_{g,t} \le P_g^{\max} - P_{g,t} \tag{4.12g}$$

$$\sum_{g,t} P_{g,t} \le L_t \tag{4.12h}$$

In order to solve the PBDED problem formulated in (4.12), the energy prices (λ_t^e) and reserve prices (λ_t^r) in (4.12a) are needed to be known. The problem is usually solved in a day ahead time horizon. Various prediction techniques exist in literature to forecast these numbers based on historical data [2]. The energy and reserve prices vs. time in price-based DED are depicted in Fig. 4.8. It should be noted that the numbers shown in Fig. 4.8 are subject to uncertainties. The GAMS code for solving the price-based DED with arbitrage is provided in GCode 4.6.

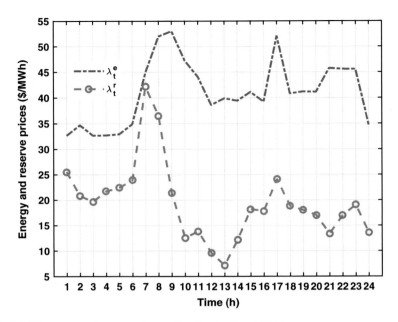

Fig. 4.8 The energy and reserve prices vs time in price-based DED

GCode 4.6 The code for price-based DED with arbitrage

```
Sets      t /t1*t24/, i /g1*g4/;
Scalar lim /inf/;
table gendata(i,*) generator cost characteristics and limits
        a     b      c    d     e     f   Pmin Pmax  RU0   RD0
g1    0.12  14.80  89   1.2  -5     3    28   200   40    40
g2    0.17  16.57  83   2.3  -4.24 6.09 20   290   30    30
g3    0.15  15.55  100  1.1  -2.15 5.69 30   190   30    30
g4    0.19  16.21  70   1.1  -3.99 6.2  20   260   50    50;
table data(t,*)
            lambda          load         Lr
t1          32.71           510.00       25.55
t2          34.72           530.00       20.83
t3          32.71           516.00       19.68
t4          32.74           510.00       21.73
t5          32.96           515.00       22.43
t6          34.93           544.00       23.94
t7          44.90           646.00       42.22
t8          52.00           686.00       36.53
t9          53.03           741.00       21.41
t10         47.26           734.00       12.58
t11         44.07           748.00       13.86
t12         38.63           760.00       9.58
t13         39.91           754.00       7.18
t14         39.45           700.00       12.16
```

t15	41.14	686.00	18.20
t16	39.23	720.00	17.83
t17	52.12	714.00	24.13
t18	40.85	761.00	18.80
t19	41.20	727.00	18.02
t20	41.15	714.00	17.03
t21	45.76	618.00	13.37
t22	45.59	584.00	16.94
t23	45.56	578.00	19.05
t24	34.72	544.00	13.58 ;

```
Variables           OF              Objective  (revenue)
                    costThermal     Cost of thermal units
                    p(i,t)          Power generated by thermal power
                                    plant
                    EM              Emission calculation
                    SR(i,t)         Spinning reserve calculation        ;
p.up(i,t) = gendata(i,"Pmax") ;
p.lo(i,t) = gendata(i,"Pmin");
SR.up(i,t)= gendata(i,"Pmax") ;
SR.lo(i,t)= 0  ;
Equations
Genconst3 , Genconst4 , costThermalcalc , balance , EMcalc , EMlim ,
      benefitcalc , reserve ;
costThermalcalc ..  costThermal =e=sum((t,i), gendata(i,'a')*power
(p(i,t),2)+gendata(i,'b')*p(i,t) +gendata(i,'c'));
Genconst3(i,t) ..  p(i,t+1)-p(i,t)=l=gendata(i,'RU0');
Genconst4(i,t) ..  p(i,t-1)-p(i,t)=l=gendata(i,'RD0');
balance(t) ..       sum(i,p(i,t))=l=data(t,'load');
EMcalc .. EM=e=sum((t,i), gendata(i,'d')*power(p(i,t),2)+gendata
(i,'e')*p(i,t)+gendata(i,'f'));
EMlim   .. EM =l=lim;
benefitcalc ..      OF=e=sum((i,t),1* data(t,'lambda')*p(i,t)+data
(t,'Lr')*SR(i,t))-costThermal;
reserve(i,t) ..     SR(i,t)=l=gendata(i,"Pmax")-p(i,t);
Model DEDPB / all /;
Solve DEDPB us qcp max of;
execute_unload "DEDPB.gdx" P.l
execute 'gdxxrw.exe DEDPB.gdx var=P   rng=P!a1 '
execute_unload "DEDPB.gdx" SR.l
execute 'gdxxrw.exe DEDPB.gdx var=SR   rng=SR!a1 '
```

The energy and reserve schedules vs time in price-based DED with arbitrage are
shown in Fig. 4.9.

4.3 Linearized Cost-Based DED

In this section, we are trying to generate a linear version of the cost-based DED
problem formulated in (4.5). The assumption is that the fuel cost function is still
expressed in quadratic form $aP^2 + bP + c$. The procedure for linearization is
illustrated in Fig. 4.10.

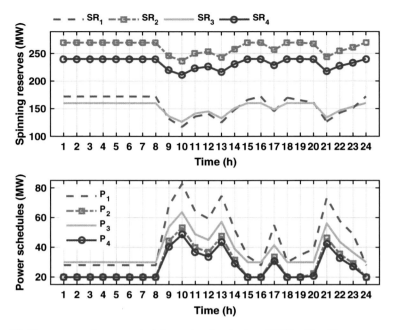

Fig. 4.9 The energy and reserve schedules vs time in price-based DED with arbitrage

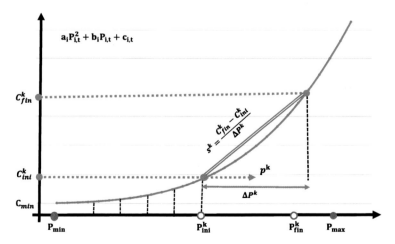

Fig. 4.10 Linearizing the quadratic cost function

As it is shown in Fig. 4.10, the interval between $\left[P_g^{\min}, P_g^{\max}\right]$ is divided into n equally sized intervals $\left[P_{g,\text{ini}}^k, P_{g,\text{fin}}^k\right]$. At a specific interval k, we define a variable $p_{g,t}^k$ which varies between zero and the interval length ΔP_g^k as indicated in (4.13b). The actual value of power generation i at time t is calculated in (4.13f). The slope of the line connecting the initial and final points of the interval k is calculated in (4.13h).

$$\min_{p_{g,t}^k} \text{OF} = \sum_{g,t} C_{g,t} \tag{4.13a}$$

$$0 \leq p_{g,t}^k \leq \Delta P_g^k, \forall k = 1 : n \tag{4.13b}$$

$$\Delta P_g^k = \frac{P_g^{\max} - P_g^{\min}}{n} \tag{4.13c}$$

$$P_{g,\text{ini}}^k = (k-1)\Delta P_g^k + P_g^{\min} \tag{4.13d}$$

$$P_{g,\text{fin}}^k = \Delta P_g^k + P_{g,\text{ini}}^k \tag{4.13e}$$

$$P_{g,t} = P_g^{\min} + \sum_k p_{g,t}^k \tag{4.13f}$$

$$C_{g,t} = a_g(P_g^{\min})^2 + b_g P_g^{\min} + c_g + \sum_k s_g^k p_{g,t}^k \tag{4.13g}$$

$$s_g^k = \frac{C_{g,\text{fin}}^k - C_{g,\text{ini}}^k}{\Delta P_g^k} \tag{4.13h}$$

$$C_{g,\text{ini}}^k = a_g(P_{g,\text{ini}}^k)^2 + b_g P_{g,\text{ini}}^k + c_g \tag{4.13i}$$

$$C_{g,\text{fin}}^k = a_g(P_{g,\text{fin}}^k)^2 + b_g P_{g,\text{fin}}^k + c_g \tag{4.13j}$$

$$P_{g,t} - P_{g,t-1} \leq \text{RU}_g \tag{4.13k}$$

$$P_{g,t-1} - P_{g,t} \leq \text{RD}_g \tag{4.13l}$$

$$\sum_{g,t} P_{g,t} \geq L_t \tag{4.13m}$$

The GAMS code for linear cost-based DED with no arbitrage in example (4.13) is provided in GCode 4.7. The $\left[P_g^{\min}, P_g^{\max}\right]$ interval is divided into 100 sections. The $\Delta P_g^k, P_{g,\text{fin}}^k, P_{g,\text{ini}}^k, C_{g,\text{fin}}^k, C_{g,\text{ini}}^k$ and s_g^k are treated as parameters.

GCode 4.7 The code for linear cost-based DED with no arbitrage in example (4.13)

```
Sets        t          hours              /t1*t24/
            i          thermal  units     /g1*g4/
            k                             /1*100/;
table  gendata(i,*)  generator  cost  characteristics  and  limits
        a      b      c     d      e      f     Pmin  Pmax  RU0   RD0
g1    0.12  14.80   89   1.2    −5      3     28   200    40    40
g2    0.17  16.57   83   2.3    −4.24  6.09  20   290    30    30
g3    0.15  15.55  100   1.1    −2.15  5.69  30   190    30    30
g4    0.19  16.21   70   1.1    −3.99  6.2   20   260    50    50;
parameter  demand(t)
/t1         510
t2          530
t3          516
t4          510
t5          515
t6          544
t7          646
t8          686
t9          741
t10         734
t11         748
t12         760
t13         754
t14         700
t15         686
t16         720
t17         714
t18         761
t19         727
t20         714
t21         618
t22         584
t23         578
t24         544/;
Variables         OF           Objective (revenue)
                  p(i,t)       Power  generated  by  thermal  power
                               plant
                  Pk(i,t,k)               ;
Parameter data(k,i,*);
data(k,i,'DP')=(gendata(i,"Pmax")−gendata(i,"Pmin"))/card(k);
data(k,i,'Pini')= (ord(k)−1)*data(k,i,'DP')+gendata(i,"Pmin");
data(k,i,'Pfin')=data(k,i,'Pini')+data(k,i,'DP');
data(k,i,'Cini')=gendata(i,"a")*power(data(k,i,'Pini'),2)
+gendata(i,"b")*data(k,i,'Pini')+gendata(i,"c");
data(k,i,'Cfin')=gendata(i,"a")*power(data(k,i,'Pfin'),2)
+gendata(i,"b")*data(k,i,'Pfin')+gendata(i,"c");
data(k,i,'s')=(data(k,i,'Cfin')−data(k,i,'Cini'))/data(k,i,'DP');
p.up(i,t)  =  gendata(i,"Pmax");
p.lo(i,t)  =  gendata(i,"Pmin");
Pk.up(i,t,k)=data(k,i,'DP');
Pk.lo(i,t,k)=0;
Equations  eq1,eq2,eq3,eq4,eq5;
```

```
eq1          .. OF =e=sum((t,i), gendata(i,'a')*power(gendata(i,"
    Pmin"),2)
+gendata(i,'b')*gendata(i,"Pmin") +gendata(i,'c')
+  sum(k,data(k,i,'s')*pk(i,t,k))   );
eq2(i,t)     .. p(i,t+1)−p(i,t)=l=gendata(i,'RU0');
eq3(i,t)     .. p(i,t−1)−p(i,t)=l=gendata(i,'RD0');
eq4(t)       .. sum(i,p(i,t))=g=demand(t);
eq5(i,t)     .. p(i,t)=e= gendata(i,"Pmin")+ sum(k,Pk(i,t,k));
Model DEDLP /all/;
Solve DEDLP us LP min OF;
```

The total cost is \$647,972.4 and compared to the operating costs obtained by GCode 4.1 which gives \$647,960.1, the error is not significant. If more precision is needed, the number of intervals (k) should be increased. This would increase the number of variables in optimization problem and makes it more difficult to solve.

4.4 Applications

There are still some other technical constraints that should be considered in DED problem formulation such as valve-point effect, prohibited operation zones (POZs), and transmission losses. References [3] propose the imperialist competitive algorithm (ICA) to solve such complicated problem. Stochastic Real-Time Scheduling of Wind-Thermal Generation Units in an Electric Utility [4].

Nomenclature

Indices

k Blocks considered for piecewise linear fuel cost function.
i Thermal generating units.
t Time intervals.

Parameters

U/S_g^0 Duration of period that unit i has been on/off at the beginning of the operating horizon (end of $t = 0$) (h).

d_g, e_g, f_g Emission coefficients of unit i.

λ_t^e Electric energy price at time t (\$/MW h).

L_t Electric demand at time t.

a_g, b_g, c_g Fuel cost coefficients of unit i.

$P_{g,\text{ini/fin}}^k$ Initial and final values of power in block k of linearized cost of thermal unit i (MW).

$C_{g,\text{ini/fin}}^k$ Initial and final values of cost in block k of linearized cost of thermal unit i (\$/h).

ΔP_g^k	Length of block k of linearized cost of thermal unit i (MW).
$P_g^{max/min}$	Maximum/minimum limits of power generation of thermal unit i.
DT/UT_g	Minimum down/up time of unit i (h).
T	Number of time intervals.
n	Number of blocks considered for piecewise linear fuel cost function.
ϵ	Parameter used for multi-objective optimization.
RU_g/RD_g	Ramp-up/down limit of generation unit i (MW/h).
λ_t^r	Reserve price at time t ($/MW).
SDC_g/SUC_g	Shut down/Start-up cost of unit i ($/h).
SD/SU_g	Shut-down/start-up ramp limit of unit i (MW/h).
s_g^k	Slope of cost in block k of linearized cost of thermal unit i ($/MW).
VWC	Value of wind curtailment
Λ_t^w	Wind availability at time t

Variables

P_t^w	Wind generation at time t
P_t^{wc}	Wind curtailment at time t
$C_{g,t}$	Fuel cost of thermal unit i at time t ($).
$p_{g,t}^k$	Operating schedule of thermal unit i in block k at time t (MW).
$P_{g,t}$	Power generated by thermal unit i at time t (MW).
$SR_{g,t}$	Spinning reserve provided by thermal unit i at time t (MW).
TC	Total operating costs ($).
EM	Total emissions (kg).

References

1. A. Soroudi, M. Afrasiab, Binary PSO-based dynamic multi-objective model for distributed generation planning under uncertainty. IET Renew. Power Gener. **6**(2), 67–78 (2012)
2. A.J. Conejo, M.A. Plazas, R. Espinola, A.B. Molina, Day-ahead electricity price forecasting using the wavelet transform and ARIMA models. IEEE Trans. Power Syst. **20**(2), 1035–1042 (2005)
3. R. Roche, L. Idoumghar, B. Blunier, A. Miraoui, Imperialist competitive algorithm for dynamic optimization of economic dispatch in power systems, in *International Conference on Artificial Evolution (Evolution Artificielle)* (Springer, Berlin, 2011), pp. 217–228
4. A. Soroudi, A. Rabiee, A. Keane, Stochastic real-time scheduling of wind-thermal generation units in an electric utility. IEEE Syst. J. **PP**(99), 1–10 (2015)

Chapter 5
Unit Commitment

This chapter provides GAMS code for solving unit commitment (UC) problem. The developed GAMS code is linear and is categorized as a MIP model in GAMS. The inputs are generator's characteristics, electricity prices, and demands. The outputs of this code are on/off status of units and their operating schedules.

Every unit commitment problem has three main cost components namely fuel costs, start-up, and shut-down costs. The unit commitment cost calculation is illustrated in Fig. 5.1.

The unit commitment data for ten units are inspired by Ademovic et al. [1] and described in Table 5.1.

5.1 Cost-Based Unit Commitment

5.1.1 Cost Calculation

$$\min_{p_{i,t}^k, u_{i,t}, y_{i,t}, z_{i,t}} \text{OF} = \sum_{i,t} \text{FC}_{i,t} + \text{STC}_{i,t} + \text{SDC}_{i,t} \tag{5.1}$$

The linear version of fuel cost calculation is described in (5.2).

$$0 \leq p_{i,t}^k \leq \Delta P_i^k u_{i,t}, \forall k = 1 : n \tag{5.2a}$$

$$\Delta P_i^k = \frac{P_i^{\max} - P_i^{\min}}{n} \tag{5.2b}$$

$$P_{i,\text{ini}}^k = (k-1)\Delta P_i^k + P_i^{\min} \tag{5.2c}$$

$$P_{i,\text{fin}}^k = \Delta P_i^k + P_{i,\text{ini}}^k \tag{5.2d}$$

© Springer International Publishing AG 2017
A. Soroudi, *Power System Optimization Modeling in GAMS*,
DOI 10.1007/978-3-319-62350-4_5

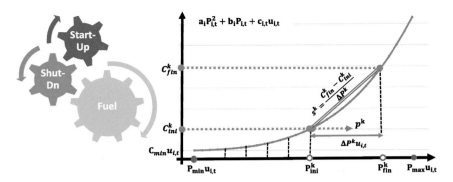

Fig. 5.1 Unit commitment cost calculation

$$P_{i,t} = P_i^{\min} u_{i,t} + \sum_k p_{i,t}^k \tag{5.2e}$$

$$C_{i,\text{ini}}^k = a_i \left(P_{i,\text{ini}}^k\right)^2 + b_i P_{i,\text{ini}}^k + c_i \tag{5.2f}$$

$$C_{i,\text{fin}}^k = a_i \left(P_{i,\text{fin}}^k\right)^2 + b_i P_{i,\text{fin}}^k + c_i \tag{5.2g}$$

$$s_i^k = \frac{C_{i,\text{fin}}^k - C_{i,\text{ini}}^k}{\Delta P_i^k} \tag{5.2h}$$

$$\text{FC}_{i,t} = a_i \left(P_i^{\min}\right)^2 + b_i P_i^{\min} + c_i u_{i,t} + \sum_k s_i^k p_{i,t}^k \tag{5.2i}$$

$u_{i,t}$ is the on/off status of the unit i at time t.

5.1.2 Ramp Rate Constraints

The power generation of unit i at time t should be within the operating limits as given in (5.3a). $\underline{P}_{i,t}$ and $\bar{P}_{i,t}$ state the minimum and maximum time-dependent operating limits. These are not necessarily equal to $P_i^{\min}$ and $P_i^{\max}$, respectively.

The upper operating limit ($\bar{P}_{i,t}$) is described in (5.3b) and (5.3c) according to [2]. The ramp up/down constraints are modeled as follows:

$$\underline{P}_{i,t} \leq P_{i,t} \leq \bar{P}_{i,t} \tag{5.3a}$$

$$\bar{P}_{i,t} \leq P_i^{\max} \left[u_{i,t} - z_{i,t+1}\right] + \text{SD}_i z_{i,t+1} \tag{5.3b}$$

$$\bar{P}_{i,t} \leq P_{i,t-1} + \text{RU}_i u_{i,t-1} + \text{SU}_i y_{i,t} \tag{5.3c}$$

$$\underline{P}_{i,t} \geq P_{i,t}^{\min} u_{i,t} \tag{5.3d}$$

$$\underline{P}_{i,t} \geq P_{i,t-1} - \text{RD}_i u_{i,t} - \text{SD}_i z_{i,t} \tag{5.3e}$$

Table 5.1 Unit commitment data for ten thermal units

Unit	a_i ($/MW^2)	b_i ($/MW)	c_i ($)	Cd_i ($)	Cs_i ($)	RU_i (MW h^{-1})	RD_i (MW h^{-1})	UT_i (h)	DT_i (h)	SD_i (MW h^{-1})	SU_i (MW h^{-1})	p_i^{min} (MW)	p_i^{max} (MW)	U_i^0 (h)	$u_{i,t=0}$	S_i^0 (h)
g1	0.0148	12.1	82	42.6	42.6	40	40	3	2	90	110	80	200	1	0	1
g2	0.0289	12.6	49	50.6	50.6	64	64	4	2	130	140	120	320	2	0	0
g3	0.0135	13.2	100	57.1	57.1	30	30	3	2	70	80	50	150	3	0	3
g4	0.0127	13.9	105	47.1	47.9	104	104	5	3	240	250	250	520	1	1	0
g5	0.0261	13.5	72	56.6	56.9	56	56	4	2	110	130	80	280	1	1	0
g6	0.0212	15.4	29	141.5	141.5	30	30	3	2	60	80	50	150	0	0	0
g7	0.0382	14	32	113.5	113.5	24	24	3	2	50	60	30	120	0	1	0
g8	0.0393	13.5	40	42.6	42.6	22	22	3	2	45	55	30	110	0	0	0
g9	0.0396	15	25	50.6	50.6	16	16	0	0	35	45	20	80	0	0	0
g10	0.0510	14.3	15	57.1	57.1	12	12	0	0	30	40	20	60	0	0	0

In order to explain the upper operating limits described in (5.3) some assumptions should be taken into account:

- It should be always less than capacity of unit i, this means that $\bar{P}_{i,t} \leq P_i^{\max}$.
- In case of unit shut-down in the next hour $(t+1)$: $\bar{P}_{i,t} \leq SD_i z_{i,t+1}$. Since $P_{i,t+1} = 0$ so $P_{i,t}$ cannot be more than SD_i.
- If the unit has been on in the previous hour $(u_{i,t-1} = 1)$ and is going to remain on then $P_{i,t}$ cannot be increased more than RU_i. This means that $\bar{P}_{i,t} \leq P_{i,t-1} + RU_i u_{i,t-1}$.
- If the unit has been off in the previous hour $(u_{i,t-1} = 0)$ and it is turned on at time t $(y_{i,t} = 1)$ then $P_{i,t}$ cannot be more than SU_i. This means that $\bar{P}_{i,t} \leq SU_i y_{i,t}$.

The combination of all these cases is enforced by (5.3b) and (5.3c).

In order to explain the lower operating limits described in (5.3) some assumptions should be taken into account:

- If the unit is on at time t then the generated power should be greater than $P_{i,t}^{\min} u_{i,t}$.
- If the unit is on at time $t-1$ and remains to be on at time $t, t+1$ then the generated power at time t should be greater than $P_{i,t-1} - RD_i u_{i,t}$.
- If the unit is on at time $t-1$ and turned off at time t then the generated power at time $t-1$ should be $P_{i,t-1} \leq SD_i z_{i,t}$.

The combination of all these cases is enforced by (5.3d) and (5.3e).

5.1.3 Min Up/Down Time Constraints

The on/off states of unit i at time t are described by $u_{i,t}$. Additionally, start-up/shut-down are given by $y_{i,t}, z_{i,t}$ in (5.4).

$$y_{i,t} - z_{i,t} = u_{i,t} - u_{i,t-1} \tag{5.4a}$$

$$y_{i,t} + z_{i,t} \leq 1 \tag{5.4b}$$

$$y_{i,t}, z_{i,t}, u_{i,t} \in \{0, 1\}$$

The minimum up time (UT_i) of unit i is modeled in (5.5) as proposed in [2].

$$\sum_{t=1}^{\zeta_i} 1 - u_{i,t} = 0 \tag{5.5a}$$

$$\sum_{t=k}^{k+UT_i-1} u_{i,t} \geq UT_i y_{i,k}, \forall k = \zeta_i + 1 \ldots T - UT_i + 1 \tag{5.5b}$$

$$\sum_{t=k}^{T} u_{i,t} - y_{i,t} \geq 0, \forall k = T - UT_i + 2 \ldots T \tag{5.5c}$$

$$\zeta_i = \min \left\{ T, \left(UT_i - U_i^0 \right) u_{i,t=0} \right\} \tag{5.5d}$$

The minimum up time (DT_i) of unit i is modeled in (5.6) as proposed in [2].

$$\sum_{t=1}^{\xi_i} u_{i,t} = 0 \tag{5.6a}$$

$$\sum_{t=k}^{k+DT_i-1} 1 - u_{i,t} \geq DT_i z_{i,k}, \forall k = \xi_i + 1 \ldots T - DT_i + 1 \tag{5.6b}$$

$$\sum_{t=k}^{T} 1 - u_{i,t} - z_{i,t} \geq 0, \forall k = T - DT_i + 2 \ldots T \tag{5.6c}$$

$$\xi_i = \min \left\{ T, \left(DT_i - S_i^0 \right) [1 - u_{i,t=0}] \right\} \tag{5.6d}$$

The realistic start-up and shut-down costs depend on how long the unit has been off or on, receptively. Some references have provided the detailed formulation for modeling these cost terms. Here for simplicity, the start-up (Sd_i) and shut-down (Sd_i) costs are modeled as constant values as (5.7).

$$STC_{i,t} = Cs_i y_{i,t} \tag{5.7a}$$

$$SDC_{i,t} = Sd_i z_{i,t} \tag{5.7b}$$

5.1.4 Demand-Generation Balance

In cost-based UC, the hourly total generation should be equal to the hourly demand:

$$\sum_{i} P_{i,t} \geq L_t \tag{5.8}$$

The hourly demand and price values vs time are depicted in Fig. 5.2.
The GAMS code for solving the cost-based unit commitment is provided in GCode 5.1.

GCode 5.1 Cost-based unit commitment example for ten unit system

```
Sets  t     time /t1*t24/,
            i generators / g1*g10 /,
                        k cost segments /sg1*sg20/,
            char /ch1*ch2/;
Alias (t,h);
Table gendata(i,*) generator cost characteristics and limits
        a    b    c   CostsD  costst  RU   RD  UT DT SD    SU    Pmin Pmax U0 Uini S0

g1   0.014 12.1  82   42.6    42.6    40   40   3  2  90   110   80   200  1  0    1
g2   0.028 12.6  49   50.6    50.6    64   64   4  2  130  140   120  320  2  0    0
g3   0.013 13.2  100  57.1    57.1    30   30   3  2  70   80    50   150  3  0    3
g4   0.012 13.9  105  47.1    47.9    104  104  5  3  240  250   250  520  1  1    0
g5   0.026 13.5  72   56.6    56.9    56   56   4  2  110  130   80   280  1  1    0
```

```
g6   0.021  15.4   29   141.5   141.5    30  30  3  2  60  80  50  150  0  0    0
g7   0.038  14.0   32   113.5   113.5    24  24  3  2  50  60  30  120  0  1    0
g8   0.039  13.5   40    42.6    42.6    22  22  3  2  45  55  30  110  0  0    0
g9   0.039  15.0   25    50.6    50.6    16  16  0  0  35  45  20   80  0  0    0
g10  0.051  14.3   15    57.1    57.1    12  12  0  0  30  40  20   60  0  0    0;
Parameter data(k,i,*);
data(k,i,'DP')=(gendata(i,"Pmax")-gendata(i,"Pmin"))/card(k);
data(k,i,'Pini')= (ord(k)-1)*data(k,i,'DP')+gendata(i,"Pmin");
data(k,i,'Pfin')=data(k,i,'Pini')+data(k,i,'DP');
data(k,i,'Cini')=gendata(i,"a")*power(data(k,i,'Pini'),2)
+gendata(i,"b")*data(k,i,'Pini')+gendata(i,"c");
data(k,i,'Cfin')=gendata(i,"a")*power(data(k,i,'Pfin'),2)
+gendata(i,"b")*data(k,i,'Pfin')+gendata(i,"c");
data(k,i,'s')= (data(k,i,'Cfin')-data(k,i,'Cini'))/data(k,i,'DP');
gendata(i,'Mincost')=gendata(i,'a')*power(gendata(i,"Pmin"),2)
+gendata(i,'b')*gendata(i,"Pmin")+gendata(i,'c');
Table dataLP(t,*)
          lambda         load
t1        14.72          883
t2        15.62          915
t23       20.50          915
t24       15.62          834;
Parameter unit(i,char);
unit(i,'ch1')=24;
unit(i,'ch2')=(gendata(i,'UT')-gendata(i,'U0'))*gendata(i,'Uini');
parameter unit2(i,char);  unit2(i,'ch1')=24;
unit2(i,'ch2')=(gendata(i,'DT')-gendata(i,'S0'))*(1-gendata(i,'Uini'));
gendata(i,'Lj')=smin(char,unit(i,char)); gendata(i,'Fj')=smin(char,unit2(i,char));
variable costThermal; positive variables pu(i,t),p(i,t),StC(i,t),SDC(i,t),Pk(i,t,k);
Binary variable u(i,t),y(i,t),z(i,t);
p.up(i,t) = gendata(i,"Pmax");  p.lo(i,t) = 0;  Pk.up(i,t,k)=data(k,i,'DP');
Pk.lo(i,t,k)=0; p.up(i,t) = gendata(i,"Pmax"); pu.up(i,h) = gendata(i,"Pmax") ;
Equations Uptime1,Uptime2,Uptime3,Dntime1,Dntime2,Dntime3,Ramp1,Ramp2,
  Ramp3,Ramp4,startc,shtdnc,genconst1,Genconst2,Genconst3,Genconst4,balance;
Uptime1(i)$(gendata(i,"Lj")>0) ..
sum(t$(ord(t)< (gendata(i,"Lj")+1)),1-U(i,t))=e=0;
Uptime2(i)$(gendata(i,"UT")>1) ..
sum(t$(ord(t)>24-gendata(i,"UT")+1),U(i,t)-y(i,t))=g=0;
Uptime3(i,t)$(ord(t)>gendata(i,"Lj") and ord(t)<24-gendata(i,"UT")+2 and
not(gendata(i,"Lj")>24-gendata(i,"UT"))) .. sum(h$((ord(h)>ord(t)-1) and
(ord(h)<ord(t)+gendata(i,"UT")),U(i,h)) =g=gendata(i,"UT")*y(i,t);
Dntime1(i)$(gendata(i,"Fj")>0) .. sum(t$(ord(t)< (gendata(i,"Fj")+1)),U(i,t))=e=0;
Dntime2(i)$(gendata(i,"DT")>1) ..
sum(t$(ord(t)>24-gendata(i,"DT")+1),1-U(i,t)-z(i,t))=g=0;
Dntime3(i,t)$(ord(t)>gendata(i,"Fj") and ord(t)<24-gendata(i,"DT")+2
and not(gendata(i,"Fj")>24-gendata(i,"DT"))   ) .. sum(h$((ord(h)>ord(t)-1)
and (ord(h)<ord(t)+gendata(i,"DT"))),1-U(i,h))=g=gendata(i,"DT")*z(i,t);
startc(i,t) .. StC(i,t)=g=gendata(i,"costst")*y(i,t);
shtdnc(i,t) .. SDC(i,t)=g=gendata(i,"CostsD")*z(i,t);
Genconst1(i,h)              .. p(i,h)=e=u(i,h)*gendata(i,"Pmin")+sum(k,Pk(i,h,k));
Genconst2(i,h)$(ord(h)>0)..U(i,h)=e=U(i,h-1)$(ord(h)>1)+gendata(i,"Uini")$(ord(h)
    =1)
                              +y(i,h)-z(i,h);
Genconst3(i,t,k)           .. Pk(i,t,k)=l=U(i,t)*data(k,i,'DP');
Genconst4 .. costThermal=e=sum((i,t),StC(i,t)+SDC(i,t))+sum((t,i),
u(i,t)*gendata(i,'Mincost')+sum(k,data(k,i,'s')*pk(i,t,k)));
Ramp1(i,t) .. p(i,t-1)-p(i,t)=l=U(i,t)*gendata(i,"RD')+z(i,t)*gendata(i,"SD");
Ramp2(i,t) .. p(i,t)=l=pu(i,t);
Ramp3(i,t)$(ord(t)<24).. pu(i,t)=l=(u(i,t)-z(i,t+1))*gendata(i,"Pmax")
                                    +z(i,t+1)*gendata(i,"SD");
Ramp4(i,t)$(ord(t)>1).. pu(i,t)=l=p(i,t-1)+U(i,t-1)*gendata(i,'RU')
                                    +y(i,t)*gendata(i,"SU");
Balance(t)..          sum(i,p(i,t))=e= dataLP(t,'load');
Model UCLP /all/;
Option optcr=0.0;
Solve UCLP minimizing costThermal using mip ;
```

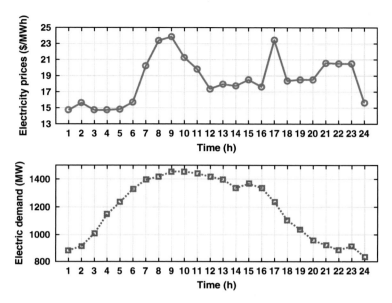

Fig. 5.2 Hourly demand and price values vs time

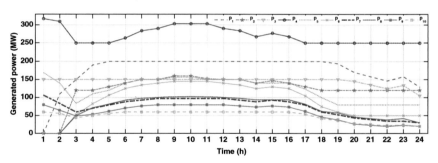

Fig. 5.3 Power schedules in cost-based unit commitment vs time

The total operating costs are \$485,240.189. The start-up, shut-down, and fuel costs are \$442.100, \$0, and \$484,798.089, respectively. The power schedules of thermal units in cost-based unit commitment are depicted in Fig. 5.3.

5.2 Cost-Based UC with Additional Constraints

In order to improve the security and efficiency of the energy supply, some additional constraints are needed to be considered in UC formulation. Some of these constraints are discussed and formulated in this section.

5.2.1   Cost-Based UC with Reserve Constraints

One of the most important resources which is used by the system operator is called spinning reserve (SR) [3]. It is used to cover the sudden increase in demand, rapid reduction of renewable energy production, or unplanned generating unit outage. The SR is supplied by online generating units which are synchronized to the system and are able to ramp-up in order to meet the demand. The UC-reserve constrained formulation is provided as follows:

$$\min_{p_{i,t}^k, u_{i,t}, y_{i,t}, z_{i,t}} \quad \text{OF} = \sum_{i,t} \text{FC}_{i,t} + \text{STC}_{i,t} + \text{SDC}_{i,t} \tag{5.9a}$$

$$\sum_i P_{i,t} \geq L_t \tag{5.9b}$$

$$R_{i,t} \leq \bar{P}_{i,t} - P_{i,t} \tag{5.9c}$$

$$\sum_i R_{i,t} \geq \gamma L_t \tag{5.9d}$$

Subject to:

(5.2), (5.3), (5.4), (5.5), (5.6), (5.7)

where $R_{i,t}$ is the reserve provided by online unit i at time t. γ is the percentage of demand which specifies the reserve requirement. It is usually expressed as a percentage of demand at time t. The GAMS code for solving the (5.9) is described in GCode 5.2. It should be noted that the generating unit characteristics are unchanged but the hourly demand values specified in GCode 5.1 and Fig. 5.2 are reduced by 55%. The γ value is specified by the system operator and is dependent on the system specification. Here, it is assumed to be 40% for simulation purpose. The total operating costs without the reserve constraint is $\$2.1039 \times 10^5$. However, if the reserve constraint is taken into account then the operating costs would increase to $\$2.1090 \times 10^5$. The operating schedules of thermal units considering reserve constraint are provided in Table 5.2.

GCode 5.2 Cost-based unit commitment with reserve constraint

```
Sets t /t1*t24/, i /g1*g10/, k /sg1*sg20/,char /ch1*ch2/; alias(t,h); alias(i,g);
Table gendata(i,*)  generator  cost  characteristics  and  limits
* Removed for saving space ;
Parameter data(k,i,*);
data(k,i,'DP')=(gendata(i,"Pmax")-gendata(i,"Pmin"))/card(k);
data(k,i,'Pini')= (ord(k)-1)*data(k,i,'DP')+gendata(i,"Pmin");
data(k,i,'Pfin')=data(k,i,'Pini')+data(k,i,'DP');
data(k,i,'Cini')=gendata(i,"a")*power(data(k,i,'Pini'),2)
+gendata(i,"b")*data(k,i,'Pini')+gendata(i,"c");
```

```
data(k,i,'Cfin')=gendata(i,"a")*power(data(k,i,'Pfin'),2)
+gendata(i,"b")*data(k,i,'Pfin')+gendata(i,"c");
data(k,i,'s')= (data(k,i,'Cfin')-data(k,i,'Cini'))/data(k,i,'DP');
gendata(i,'Mincost')=gendata(i,"a")*power(gendata(i,"Pmin"),2)
+gendata(i,'b')*gendata(i,"Pmin")+gendata(i,'c');
table dataLP(t,*)
                lambda         load
* Removed for saving space       ;
dataLP(t,'load')=dataLP(t,'load')*0.45; Parameter unit(i,char);
unit(i,'ch1')=24; unit(i,'ch2')=(gendata(i,'UT')-gendata(i,'U0'))*gendata(i,'Uini')
  ;
Parameter unit2(i,char); unit2(i,'ch1')=24;
unit2(i,'ch2')=(gendata(i,'DT')-gendata(i,'S0'))*(1-gendata(i,'Uini'));
gendata(i,'Lj')=smin(char,unit(i,char)); gendata(i,'Fj')=smin(char,unit2(i,char));
Variable  costThermal; Binary variable u(i,t),y(i,t),z(i,t);
Positive variables pu(i,t),p(i,t),StC(i,t),SDC(i,t),Pk(i,t,k);
p.up(i,t) = gendata(i,"Pmax") ; p.lo(i,t) =0;
Pk.up(i,t,k)=data(k,i,'DP'); Pk.lo(i,t,k)=0;
p.up(i,t) = gendata(i,"Pmax"); pu.up(i,h) = gendata(i,"Pmax");
Equations Uptime1 ,Uptime2 ,Uptime3 ,Dntime1 ,Dntime2 ,Dntime3 ,
Ramp1 ,Ramp2 ,Ramp3 ,Ramp4 , startc ,shtdnc ,genconst1 ,
Genconst2 ,Genconst3 ,Genconst4 , balance , reserve ;
Uptime1 (i)$(gendata(i,"Lj")>0)..
 sum(t$(ord(t)< (gendata(i,"Lj")+1)),1-U(i,t))=e=0;
Uptime2 (i)$(gendata(i,"UT")>1)..
 sum(t$(ord(t)>24-gendata(i,"UT")+1),U(i,t)-y(i,t))=g=0;
Uptime3 (i,t)$(ord(t)>gendata(i,"Lj") and ord(t)<24-gendata(i,"UT")+2 and
not(gendata(i,"Lj")>24-gendata(i,"UT")))
    .. sum(h$((ord(h)>ord(t)-1) and (ord(h)<ord(t)+gendata(i,"UT"))),U(i,h)) =g=
                      gendata(i,"UT")*y(i,t);
Dntime1 (i)$(gendata(i,"Fj")>0) ..   sum(t$(ord(t)< (gendata(i,"Fj")+1)),U(i,t))=e
  =0;
Dntime2 (i)$(gendata(i,"DT")>1) ..
                      sum(t$(ord(t)>24-gendata(i,"DT")+1),1-U(i,t)-z(i,t))=g
                      =0;
Dntime3 (i,t)$(ord(t)>gendata(i,"Fj") and ord(t)<24-gendata(i,"DT")+2 and
not(gendata(i,"Fj")>24-gendata(i,"DT")))
.. sum(h$((ord(h)>ord(t)-1) and (ord(h)<ord(t)+gendata(i,"DT"))),1-U(i,h)) =g=
gendata(i,"DT")*z(i,t);
startc(i,t) .. StC(i,t)=g=gendata(i,"costst")*y(i,t);
shtdnc(i,t) .. SDC(i,t)=g=gendata(i,"CostsD")*z(i,t);
genconst1(i,h) .. p(i,h)=e=u(i,h)*gendata(i,"Pmin")+sum(k,Pk(i,h,k));
Genconst2(i,h)$(ord(h)>0)..U(i,h)=e=
U(i,h-1)$(ord(h)>1)+gendata(i,"Uini")$(ord(h)=1)+y(i,h)-z(i,h);
Genconst3(i,t,k) .. Pk(i,t,k)=l=U(i,t)*data(k,i,'DP');
Genconst4   .. costThermal=e=sum((i,t),StC(i,t)+SDC(i,t))+sum((t,i),
                      u(i,t)*gendata(i,'Mincost')+sum(k,data(k,i,'s')*pk(i,t,k)));
Ramp1(i,t) .. p(i,t-1)-p(i,t)=l=U(i,t)*gendata(i,'RD')+z(i,t)*gendata(i,"SD");
Ramp2(i,t) .. p(i,t)=l=pu(i,t);
Ramp3(i,t)$(ord(t)<24)..pu(i,t)=l=(u(i,t)-z(i,t+1))*gendata(i,"Pmax")
                                         +z(i,t+1)*gendata(i,"SD");
Ramp4(i,t)$(ord(t)>1)  .. pu(i,t)=l=p(i,t-1)+U(i,t-1)*gendata(i,'RU')
                                         +y(i,t)*gendata(i,"SU");
balance(t) .. sum(i,p(i,t))=e= dataLP(t,'load');
reserve(t) .. sum(i,pu(i,t)-p(i,t))=g=0.40*dataLP(t,'load');
Model UCLP / all /; Option optcr=0.0;
Solve UCLP minimizing costThermal using mip ;
```

Table 5.2 Power schedules ($P_{i,t}$) of thermal units considering reserve constraint

Time	$g1$	$g3$	$g4$	$g5$	$g7$	$g8$	$g10$
t_1			250.0	90.4	57.0		
t_2			250.0	107.8	54.0		
t_3	94.5		250.0	80.0	30.0		
t_4	92.1	65.0	250.0	80.0	30.0		
t_5	104.0	72.2	250.0	80.0	30.0		20.0
t_6	110.0	79.0	250.0	80.0	30.0	30.0	20.0
t_7	123.7	95.0	250.0	80.0	30.0	30.0	20.0
t_8	128.0	100.6	250.0	80.0	30.0	30.0	20.0
t_9	134.0	106.8	250.0	80.0	30.0	34.0	20.0
t_{10}	134.0	106.8	250.0	80.0	30.0	34.0	20.0
t_{11}	133.5	105.0	250.0	80.0	30.0	30.0	20.0
t_{12}	134.6	110.0	250.0	80.0	30.0	34.0	
t_{13}	133.7	105.0	250.0	80.0	30.0	30.0	
t_{14}	122.0	90.6	250.0	80.0	30.0	30.0	
t_{15}	128.0	97.6	250.0	80.0	30.0	30.0	
t_{16}	122.0	90.6	250.0	80.0	30.0	30.0	
t_{17}	98.0	68.2	250.0	80.0	30.0	30.0	
t_{18}	110.0	77.3	250.0		30.0	30.0	
t_{19}	104.2	52.9	250.0		30.0	30.0	
t_{20}	121.6		250.0		30.0	30.0	
t_{21}	104.9		250.0		30.0	30.0	
t_{22}	88.3		250.0		30.0	30.0	
t_{23}	101.8		250.0		30.0	30.0	
t_{24}	95.3		250.0		30.0		

The reserve provision ($R_{i,t}$) by thermal units considering reserve constraint is given in Table 5.3.

5.2.2 Cost-Based UC Considering Generator Contingency

In this section, the cost-based UC is formulated in order to consider the generator outage contingencies. The system operator is willing to be robust against the unexpected outages of generating units. It means that in case of any generating unit outage the reserve resource by the remaining units should be able to supply the lost generating unit. This constraint is modeled as follows:

$$\min_{p_{i,t}^k, u_{i,t}, y_{i,t}, z_{i,t}} \quad \text{OF} = \sum_{i,t} \text{FC}_{i,t} + \text{STC}_{i,t} + \text{SDC}_{i,t} \tag{5.10a}$$

Table 5.3 Reserve provision ($R_{i,t}$) by thermal units considering reserve constraint

Time	g1	g3	g4	g5	g7	g8	g10
t_1			270.0	189.7	63.0		
t_2			99.1	38.6	27.0		
t_3	15.5		34.6	83.8	48.0		
t_4	42.5	15.0	69.4	56.0	24.0		
t_5	28.1	22.8	71.6	56.0	24.0		20.0
t_6	34.0		104.0	56.0	8.6	25.0	12.0
t_7	26.4	14.0	97.2	56.0	24.0	22.0	12.0
t_8	35.7	24.5	81.3	56.0	24.0	22.0	12.0
t_9	34.0	23.8	94.1	56.0	24.0	18.0	12.0
t_{10}	40.0	30.0	77.9	56.0	24.0	22.0	12.0
t_{11}	40.6	31.8	71.1	56.0	24.0	26.0	10.0
t_{12}	38.9	25.0	93.5	56.0	24.0	18.0	
t_{13}	40.9	35.0	69.6	56.0	24.0	26.0	
t_{14}	51.7	44.5	42.9	56.0	24.0	22.0	
t_{15}	34.0	23.0	87.3	56.0	24.0	22.0	
t_{16}	46.0	37.1	56.0	56.0	24.0	22.0	
t_{17}	42.5		104.0	30.0	24.0	22.0	
t_{18}	28.0	21.0	104.0		24.0	22.0	
t_{19}	45.8	17.1	77.9		24.0	22.0	
t_{20}	22.6		104.0		24.0	22.0	
t_{21}	56.7		63.3		24.0	22.0	
t_{22}	56.7		56.7		24.0	22.0	
t_{23}	21.7		104.0		24.0	15.0	
t_{24}			98.1			52.0	

$$\sum_i P_{i,t} \geq L_t \tag{5.10b}$$

$$R_{i,t} \leq \bar{P}_{i,t} - P_{i,t} \tag{5.10c}$$

$$\sum_{i \neq i'} R_{i,t} \geq P_{i',t} \quad \forall i' \in \Omega_c \tag{5.10d}$$

Subject to:

(5.2), (5.3), (5.4), (5.5), (5.6), (5.7)

where Ω_c is the set of contingencies for generating units. The GAMS code for solving the (5.10) is described in GCode 5.3. As previously stated, the total operating costs without the generator contingency constraint are 2.1039×10^5. However, if the generator contingency constraint (the outage of all generating units except g4 is considered) is taken into account then the operating costs would increase to 2.1041×10^5.

GCode 5.3 Cost-based unit commitment considering generator contingencies

```
Sets  t      time  /t1*t24/
      i      generators  indices  /  g1*g10  /,  k    cost  segments  /sg1*sg20/
      char  /ch1*ch2/,  g(i)  /g1*g3,g5*g10/;  Alias  (t,h);
table  gendata(i,*)  generator  cost  characteristics  and  limits
      a    b    c   CostsD    costst    RU  RD  UT  DT  SD  SU  Pmin  Pmax  U0  Uini  S0
*  Removed  for  saving  space                 ;
parameter  data(k,i,*);
data(k,i,'DP')=(gendata(i,"Pmax")—gendata(i,"Pmin"))/card(k);
data(k,i,'Pini')=  (ord(k)—1)*data(k,i,'DP')+gendata(i,"Pmin");
data(k,i,'Pfin')=data(k,i,'Pini')+data(k,i,'DP');
data(k,i,'Cini')=gendata(i,"a")*power(data(k,i,'Pini'),2)
+gendata(i,"b")*data(k,i,'Pini')+gendata(i,"c");
data(k,i,'Cfin')=gendata(i,"a")*power(data(k,i,'Pfin'),2)
+gendata(i,"b")*data(k,i,'Pfin')+gendata(i,"c");
data(k,i,'s')=  (data(k,i,'Cfin')—data(k,i,'Cini'))/data(k,i,'DP');
gendata(i,'Mincost')=gendata(i,'a')*power(gendata(i,"Pmin"),2)
+gendata(i,'b')*gendata(i,"Pmin")+gendata(i,'c');
Table  dataLP(t,*)
            lambda          load
*  Removed  for  saving  space;
dataLP(t,'load')=dataLP(t,'load')*0.45;
Parameter  unit(i,char);  unit(i,'ch1')=24;
unit(i,'ch2')=(gendata(i,'UT')—gendata(i,'U0'))*gendata(i,'Uini');
parameter  unit2(i,char);  unit2(i,'ch1')=24;
unit2(i,'ch2')=(gendata(i,'DT')—gendata(i,'S0'))*(1—gendata(i,'Uini'));
gendata(i,'Lj')=smin(char,unit(i,char));gendata(i,'Fj')=smin(char,unit2(i,char));
Variable   costThermal   ;
positive  variables  pu(i,t),p(i,t),StC(i,t),SDC(i,t),Pk(i,t,k);
Binary  variable  u(i,t),y(i,t),z(i,t);
p.up(i,t)  =  gendata(i,"Pmax")  ;  p.lo(i,t)  =  0;
Pk.up(i,t,k)=data(k,i,'DP');  Pk.lo(i,t,k)=0;
p.up(i,t)  =  gendata(i,"Pmax");  pu.up(i,h)  =  gendata(i,"Pmax");
Equations  Uptime1,Uptime2,Uptime3,Dntime1,Dntime2,Dntime3,Ramp1,Ramp2,Ramp3,Ramp4
,startc,shtdnc,genconst1,Genconst2,Genconst3,Genconst4,balance,reserve;
Uptime1(i)$(gendata(i,"Lj")>0)..sum(t$(ord(t)< (gendata(i,"Lj")+1)),1—U(i,t))=e=0;
Uptime2(i)$(gendata(i,"UT")>1)
                         ..sum(t$(ord(t)>24—gendata(i,"UT")+1),U(i,t)—y(i,t))=g=0;
Uptime3(i,t)$(ord(t)>gendata(i,"Lj")  and  ord(t)<24—gendata(i,"UT")+2
and  not(gendata(i,"Lj")>24—gendata(i,"UT"))    )  ..  sum(h$((ord(h)>ord(t)—1) and
(ord(h)<ord(t)+gendata(i,"UT"))),U(i,h))  =g=gendata(i,"UT")*y(i,t);
Dntime1(i)$(gendata(i,"Fj")>0)..sum(t$(ord(t)<(gendata(i,"Fj")+1)),U(i,t))=e=0;
Dntime2(i)$(gendata(i,"DT")>1)
..  sum(t$(ord(t)>24—gendata(i,"DT")+1),1—U(i,t)—z(i,t))=g=0;
Dntime3(i,t)$(ord(t)>gendata(i,"Fj")  and  ord(t)<24—gendata(i,"DT")+2 and
not(gendata(i,"Fj")>24—gendata(i,"DT"))    )  ..  sum(h$((ord(h)>ord(t)—1) and
(ord(h)<ord(t)+gendata(i,"DT"))),1—U(i,h))  =g=gendata(i,"DT")*z(i,t);
startc(i,t)  ..  StC(i,t)=g=gendata(i,"costst")*y(i,t);
shtdnc(i,t)  ..  SDC(i,t)=g=gendata(i,"CostsD")*z(i,t);
genconst1(i,h)  ..  p(i,h)=e=u(i,h)*gendata(i,"Pmin")+sum(k,Pk(i,h,k));
Genconst2(i,h)$(ord(h)>0)  ..  U(i,h)=e=U(i,h—1)$(ord(h)>1)
+gendata(i,"Uini")$(ord(h)=1)+y(i,h)—z(i,h);
Genconst3(i,t,k)  ..  Pk(i,t,k)=1=U(i,t)*data(k,i,'DP');
Genconst4  ..costThermal=e=sum((i,t),StC(i,t)+SDC(i,t))
+sum((t,i),u(i,t)*gendata(i,'Mincost')+sum(k,data(k,i,'s')*pk(i,t,k)));
Ramp1(i,t)..  p(i,t—1)—p(i,t)=1=U(i,t)*gendata(i,'RD')+z(i,t)*gendata(i,"SD");
Ramp2(i,t)..  p(i,t)=1=pu(i,t);
Ramp3(i,t)$(ord(t)<24)..  pu(i,t)=1=(u(i,t)—z(i,t+1))*gendata(i,"Pmax")
                                    +z(i,t+1)*gendata(i,"SD");
Ramp4(i,t)$(ord(t)>1)  ..pu(i,t)=1=p(i,t—1)+U(i,t—1)*gendata(i,'RU')
                                    +y(i,t)*gendata(i,"SU");
balance(t)  ..          sum(i,p(i,t))=e=  dataLP(t,'load');
```

```
reserve(t,g) .. sum(i$(ord(i) <> ord(g) ),pu(i,t)-p(i,t))=g=1*p(g,t);
Model UCLP /all/; Option optcr=0.0;
Solve UCLP minimizing costThermal using mip ;
```

Power schedules $(P_{i,t})$ by thermal units considering generator contingency constraint are provided in Table 5.4.

The reserve provision $(R_{i,t})$ by thermal units considering generator contingencies is provided in Table 5.5.

5.2.3   Cost-Based UC with Demand Flexibility Constraints

The balance between generation and demand is traditionally managed by scheduling the generating units. However, this paradigm is changing gradually. In other words, the demand values can also change intentionally to increase the efficiency of UC

Table 5.4 Power schedules $(P_{i,t})$ by thermal units considering generator contingency constraint

Time	g1	g3	g4	g5	g7	g8	g10
t_1	0.0	0.0	250.0	93.4	54.0	0.0	0.0
t_2	0.0	51.8	250.0	80.0	30.0	0.0	0.0
t_3	0.0	74.5	250.0	80.0	30.0	0.0	20.0
t_4	110.0	104.5	250.0	0.0	30.6	0.0	22.0
t_5	128.0	98.2	250.0	0.0	30.0	30.0	20.0
t_6	146.0	115.0	250.0	0.0	30.0	38.0	20.0
t_7	152.0	130.0	250.0	0.0	34.5	40.2	22.0
t_8	158.0	130.1	250.0	0.0	34.5	42.0	24.0
t_9	162.8	135.0	250.0	0.0	39.0	42.0	26.0
t_{10}	162.8	135.0	250.0	0.0	39.0	42.0	26.0
t_{11}	158.0	135.0	250.0	0.0	39.0	42.0	24.5
t_{12}	158.0	130.1	250.0	0.0	34.5	42.0	24.0
t_{13}	152.0	130.0	250.0	0.0	34.5	40.2	22.0
t_{14}	146.0	118.6	250.0	0.0	30.0	38.0	20.0
t_{15}	146.1	125.0	250.0	0.0	34.5	38.0	22.0
t_{16}	146.0	118.6	250.0	0.0	30.0	38.0	20.0
t_{17}	139.0	113.3	250.0	0.0	0.0	34.0	20.0
t_{18}	137.3	110.0	250.0	0.0	0.0	0.0	0.0
t_{19}	122.1	95.0	250.0	0.0	0.0	0.0	0.0
t_{20}	106.6	75.0	250.0	0.0	0.0	0.0	0.0
t_{21}	98.0	66.9	250.0	0.0	0.0	0.0	0.0
t_{22}	92.0	56.3	250.0	0.0	0.0	0.0	0.0
t_{23}	96.8	65.0	250.0	0.0	0.0	0.0	0.0
t_{24}	125.3	0.0	250.0	0.0	0.0	0.0	0.0

Table 5.5 Reserve provision ($R_{i,t}$) by thermal units considering generator contingency constraint

Time	$g1$	$g3$	$g4$	$g5$	$g7$	$g8$	$g10$
t_1	0.0	0.0	270.0	186.7	66.0	0.0	0.0
t_2	0.0	28.3	0.0	69.4	48.0	0.0	0.0
t_3	0.0	7.3	1.8	30.0	22.8	0.0	20.0
t_4	0.0	0.0	76.6	0.0	23.5	0.0	10.0
t_5	22.0	0.0	64.5	0.0	24.6	25.0	14.0
t_6	0.0	4.0	104.0	0.0	24.0	14.0	0.0
t_7	0.0	8.7	104.0	0.0	19.5	19.9	0.0
t_8	0.0	0.0	103.9	0.0	24.0	20.2	10.0
t_9	0.0	15.0	96.3	0.0	19.5	22.0	10.0
t_{10}	0.0	15.0	89.8	0.0	24.0	22.0	12.0
t_{11}	0.0	0.0	98.5	0.0	24.0	22.0	13.6
t_{12}	0.0	0.0	95.1	0.0	28.5	22.0	12.5
t_{13}	0.0	0.2	104.0	0.0	24.0	23.9	0.0
t_{14}	0.0	3.9	104.0	0.0	0.0	24.2	14.0
t_{15}	0.0	21.1	73.5	0.0	19.5	22.0	10.0
t_{16}	0.0	0.0	90.0	0.0	20.0	22.0	14.0
t_{17}	0.0	14.0	104.0	0.0	0.0	11.0	10.0
t_{18}	6.0	33.3	104.0	0.0	0.0	0.0	0.0
t_{19}	0.0	18.1	104.0	0.0	0.0	0.0	0.0
t_{20}	0.0	31.6	75.0	0.0	0.0	0.0	0.0
t_{21}	0.0	31.1	66.9	0.0	0.0	0.0	0.0
t_{22}	0.0	0.0	92.0	0.0	0.0	0.0	0.0
t_{23}	0.0	5.0	91.8	0.0	0.0	0.0	0.0
t_{24}	0.0	95.0	30.3	0.0	0.0	0.0	0.0

problems. In this modern context, the demand values are no longer strict and can be dispatched. The UC problem incorporating a simple demand response model is presented as follows:

$$\min_{p_{i,t}^k, u_{i,t}, y_{i,t}, z_{i,t}} \text{OF} = \sum_{i,t} \text{FC}_{i,t} + \text{STC}_{i,t} + \text{SDC}_{i,t} \tag{5.11a}$$

$$\sum_i P_{i,t} \geq D_t \tag{5.11b}$$

$$(1 - \varsigma_{\min})L_t \leq D_t \leq (1 + \varsigma_{\max})L_t \tag{5.11c}$$

$$\sum_t D_t = \sum_t L_t \tag{5.11d}$$

Subject to:

(5.2), (5.3), (5.4), (5.5), (5.6), (5.7)

Equation (5.11c) states the demand variation ranges. $\varsigma_{\min/\max}$ determine the min/-max flexibility of demand response. Equation (5.11d) states that the total energy of the consumer does not change over the operating horizon. The $\varsigma_{\min/\max}$ are assumed to be 10%.

The GAMS code for solving the (5.11) is described in GCode 5.4. As previously stated, the total operating costs without the generator contingency constraint are 2.1039×10^5. However, if the demand response flexibility is available then the operating costs would decrease to 2.0965×10^5. The power schedules ($P_{i,t}$) by thermal units considering demand response constraint are given in Table 5.6.

GCode 5.4 Cost-based unit commitment considering demand response

```
Sets t/t1*t24/, i/g1*g10 /,k/sg1*sg20/,char /ch1*ch2/;alias (t,h);
Table gendata(i,*) generator cost characteristics and limits
          a    b    c     CostsD   costst   RU RD UT DT SD SU Pmin Pmax U0 Uini S0
* Removed for saving space      ; parameter data(k,i,*);
data(k,i,'DP')=(gendata(i,"Pmax")—gendata(i,"Pmin"))/card(k);
data(k,i,'Pini')= (ord(k)—1)*data(k,i,'DP')+gendata(i,"Pmin");
data(k,i,'Pfin')=data(k,i,'Pini')+data(k,i,'DP');
data(k,i,'Cini')=gendata(i,"a")*power(data(k,i,'Pini'),2)
        +gendata(i,"b")*data(k,i,'Pini')+gendata(i,"c");
data(k,i,'Cfin')=gendata(i,"a")*power(data(k,i,'Pfin'),2)
+gendata(i,"b")*data(k,i,'Pfin')+gendata(i,"c");
data(k,i,'s')= (data(k,i,'Cfin')—data(k,i,'Cini'))/data(k,i,'DP');
gendata(i,'Mincost')=gendata(i,'a')*power(gendata(i,"Pmin"),2)
                     +gendata(i,'b')*gendata(i,"Pmin")+gendata(i,'c');
Table dataLP(t,*)
            lambda           load
* Removed for saving space        ;
dataLP(t,'load')=dataLP(t,'load')*0.45;
parameter unit(i,char);
unit(i,'ch1')=24;  unit(i,'ch2')=
(gendata(i,'UT')—gendata(i,'U0'))*gendata(i,'Uini');
Parameter unit2(i,char);
unit2(i,'ch1')=24;unit2(i,'ch2')=
(gendata(i,'DT')—gendata(i,'S0'))*(1—gendata(i,'Uini'));
gendata(i,'Lj')=smin(char,unit(i,char));
gendata(i,'Fj')=smin(char,unit2(i,char));  Variable   costThermal    ;
Positive variables pu(i,t),p(i,t),StC(i,t),SDC(i,t),Pk(i,t,k),D(t);
Binary variable u(i,t),y(i,t),z(i,t);
p.up(i,t) = gendata(i,"Pmax")  ;
p.lo(i,t) = 0; Pk.up(i,t,k)=data(k,i,'DP'); Pk.lo(i,t,k)=0;
p.up(i,t) = gendata(i,"Pmax")  ; pu.up(i,h) = gendata(i,"Pmax")  ;
Equations Uptime1,Uptime2,Uptime3,Dntime1,Dntime2,Dntime3,Ramp1,Ramp2,Ramp3,
Ramp4,startc,shtdnc,genconst1,Genconst2,Genconst3,Genconst4,balance,DRconst;
Uptime1(i)$(gendata(i,"Lj")>0)..sum(t$(ord(t)< (gendata(i,"Lj")+1)),1—U(i,t))=e=0;
Uptime2(i)$(gendata(i,"UT")>1) ..
sum(t$(ord(t)>24—gendata(i,"UT")+1),U(i,t)—y(i,t))=g=0;
Uptime3(i,t)$(ord(t)<24—gendata(i,"UT") and ord(t)<24—gendata(i,"UT")+2
and not(gendata(i,"Lj")>24—gendata(i,"UT"))   ) .. sum(h$((ord(h)>ord(t)—1) and
(ord(h)<ord(t)+gendata(i,"UT"))),U(i,h)) =g=gendata(i,"UT")*y(i,t);
Dntime1(i)$(gendata(i,"Fj")>0) .. sum(t$(ord(t)< (gendata(i,"Fj")+1)),U(i,t))=e=0;
Dntime2(i)$(gendata(i,"DT")>1) ..
sum(t$(ord(t)>24—gendata(i,"DT")+1),1—U(i,t)—z(i,t))=g=0;
Dntime3(i,t)$(ord(t)>gendata(i,"Fj") and ord(t)<24—gendata(i,"DT")+2 and
not(gendata(i,"Fj")>24—gendata(i,"DT"))   ) .. sum(h$((ord(h)>ord(t)—1) and
(ord(h)<ord(t)+gendata(i,"DT"))),1—U(i,h)) =g=gendata(i,"DT")*z(i,t);
startc(i,t) .. StC(i,t)=g=gendata(i,"costst")*y(i,t);
shtdnc(i,t) .. SDC(i,t)=g=gendata(i,"CostsD")*z(i,t);
```

```
genconst1(i,h)              .. p(i,h)=e=u(i,h)*gendata(i,"Pmin")+sum(k,Pk(i,h,k));
Genconst2(i,h)$(ord(h)>0)  .. U(i,h)=e=U(i,h−1)$(ord(h)>1)
+gendata(i,"Uini")$(ord(h)=1)+y(i,h)−z(i,h);
Genconst3(i,t,k)          .. Pk(i,t,k)=l=U(i,t)*data(k,i,'DP');
Genconst4 .. costThermal=e=sum((i,t),StC(i,t)+SDC(i,t))
+sum((t,i),u(i,t)*gendata(i,'Mincost')+sum(k,data(k,i,'s')*pk(i,t,k)));
Ramp1(i,t) .. p(i,t−1)−p(i,t)=l=U(i,t)*gendata(i,'RD')+z(i,t)*gendata(i,"SD");
Ramp2(i,t) .. p(i,t)=l=pu(i,t);
Ramp3(i,t)$(ord(t)<24)..pu(i,t)=l=(u(i,t)−z(i,t+1))*gendata(i,"Pmax")
+z(i,t+1)*gendata(i,"SD");
Ramp4(i,t)$(ord(t)>1)  ..pu(i,t)=l=p(i,t−1)+U(i,t−1)*gendata(i,'RU')
+y(i,t)*gendata(i,"SU");
balance(t) ..      sum(i,p(i,t))=e= D(t);
DRconst     ..      sum(t,dataLP(t,'load'))=e=sum(t,D(t));
Model UCLP / all /;
Option optcr=0.0; D.up(t)=1.1*dataLP(t,'load'); D.lo(t)=0.9*dataLP(t,'load');
Solve UCLP minimizing costThermal using mip ;
```

Table 5.6 Power schedules ($P_{i,t}$) by thermal units considering demand response constraint

Time	g1	g3	g4	g5	g7	g8
t_1			250.0	80.0	34.5	
t_2	92.9		250.0	80.0	30.0	
t_3	132.9		250.0	80.0	34.5	
t_4	146.0	80.0	250.0		34.5	38.0
t_5	146.0	110.0	250.0		34.5	38.0
t_6	146.0	120.0	250.0		34.5	38.0
t_7	146.0	120.0	250.0		34.5	38.0
t_8	146.0	120.0	250.0		34.5	38.0
t_9	146.0	120.8	250.0		34.5	38.0
t_{10}	146.0	120.8	250.0		34.5	38.0
t_{11}	146.0	120.0	250.0		34.5	38.0
t_{12}	146.0	120.0	250.0		34.5	38.0
t_{13}	146.0	120.0	250.0		34.5	38.0
t_{14}	146.0	120.0	250.0		34.5	38.0
t_{15}	146.0	120.0	250.0		34.5	38.0
t_{16}	146.0	120.0	250.0		34.5	38.0
t_{17}	146.0	120.0	250.0		34.5	38.0
t_{18}	144.0	115.0	250.0			38.0
t_{19}	128.8	105.0	250.0			30.0
t_{20}	128.0	96.7	250.0			
t_{21}	116.4	90.0	250.0			
t_{22}	110.0	78.1	250.0			
t_{23}	116.0	86.9	250.0			
t_{24}	97.8	65.0	250.0			

The demand pattern change in cost-based unit commitment with demand response flexibility is shown in Fig. 5.4. The demand flexibility is changed from 0 to 18% then the demand pattern changes are shown in Fig. 5.5. The variation of total operating costs vs demand flexibility is depicted in Fig. 5.6.

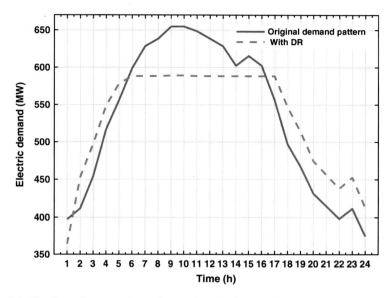

Fig. 5.4 The demand pattern change in cost-based unit commitment with demand response flexibility

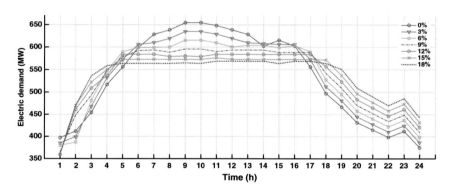

Fig. 5.5 The sensitivity analysis of demand pattern changes in CBUC with various demand response flexibilities

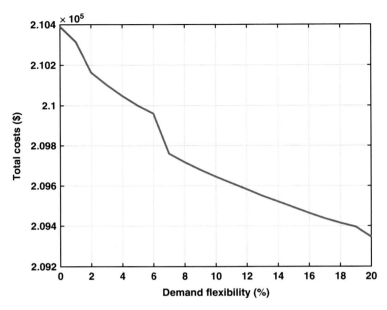

Fig. 5.6 The variation of total operating costs vs demand flexibility

5.3   Price-Based Unit Commitment

The benefit maximization in price-based unit commitment is formulated in (5.12).

$$\max_{p_{i,t}^k, u_{i,t}, y_{i,t}, z_{i,t}} \quad \text{OF} = \sum_{i,t} \{\lambda_t p_{i,t} - [\text{FC}_{i,t} + \text{STC}_{i,t} + \text{SDC}_{i,t}]\} \tag{5.12}$$

Subject to:

(5.2), (5.3), (5.4), (5.5), (5.6), (5.7)

The GAMS code for solving the (5.12) is described in GCode 5.5.

GCode 5.5 Price-based unit commitment Example for ten unit system

```
Sets  t    /t1*t24/, i  /g1*g10 /,k  /sg1*sg20/,char  /ch1*ch2/;
Alias (t,h)
Table Gdata(i,*) generator cost characteristics and limits
       a   b   c   CostsD   costst RU RD UT DT SD SU Pmin Pmax U0 Uini S0
* Removed for saving space
Parameter data(k,i,*);
data(k,i,'DP')=(Gdata(i,"Pmax")—Gdata(i,"Pmin"))/card(k);
data(k,i,'Pini')= (ord(k)—1)*data(k,i,'DP')+Gdata(i,"Pmin");
data(k,i,'Pfin')=data(k,i,'Pini')+data(k,i,'DP');
data(k,i,'Cini')=Gdata(i,"a")*power(data(k,i,'Pini'),2)
```

```
                              +Gdata(i,"b")*data(k,i,'Pini')+Gdata(i,"c");
data(k,i,'Cfin')=Gdata(i,"a")*power(data(k,i,'Pfin'),2)
                              +Gdata(i,"b")*data(k,i,'Pfin')+Gdata(i,"c");
data(k,i,'s')= (data(k,i,'Cfin')─data(k,i,'Cini'))/data(k,i,'DP');
Gdata(i,'Mincost')=Gdata(i,'a')*power(Gdata(i,"Pmin"),2)
                              +Gdata(i,'b')*Gdata(i,"Pmin") +Gdata(i,'c');
table dataLP(t,*)
          lambda           load
* Removed for saving space ;
parameter unit(i,char); unit(i,'ch1')=24;
unit(i,'ch2')=(Gdata(i,'UT')─Gdata(i,'U0'))*Gdata(i,'Uini');
Parameter unit2(i,char);
unit2(i,'ch1')=24;
unit2(i,'ch2')=(Gdata(i,'DT')─Gdata(i,'S0'))*(1─Gdata(i,'Uini'));
Gdata(i,'Lj')=smin(char,unit(i,char));
Gdata(i,'Fj')=smin(char,unit2(i,char)); Variable   costThermal ,OF;
Positive variables pu(i,t),p(i,t),StC(i,t),SDC(i,t),Pk(i,t,k);
binary variable u(i,t),y(i,t),z(i,t);
p.up(i,t) = Gdata(i,"Pmax") ; p.lo(i,t) = 0;
Pk.up(i,t,k)=data(k,i,'DP'); Pk.lo(i,t,k)=0;
p.up(i,t) = Gdata(i,"Pmax") ; pu.up(i,h) = Gdata(i,"Pmax") ;
Eequations Upt1,Upt2,Upt3,Dntime1,Dntime2,Dntime3,Ramp1,Ramp2,Ramp3,Ramp4,
startc ,shtdnc,genconst1, Genconst2,Genconst3,Genconst4,balance,benefitcalc;
Upt1(i)$(Gdata(i,"Lj")>0)..sum(t$(ord(t)< (Gdata(i,"Lj")+1)),1─U(i,t))=e=0;
Upt2(i)$(Gdata(i,"UT")>1)..sum(t$(ord(t)>24─Gdata(i,"UT")+1),U(i,t)─y(i,t))=g=0;
Upt3(i,t)$(ord(t)>Gdata(i,"Lj") and ord(t)<24─Gdata(i,"UT")+2 and
not(Gdata(i,"Lj")>24─Gdata(i,"UT"))) ..sum(h$((ord(h)>ord(t)─1) and
(ord(h)<ord(t)+Gdata(i,"UT"))),U(i,h)) =g=Gdata(i,"UT")*y(i,t);
Dntime1(i)$(Gdata(i,"Fj")>0)..sum(t$(ord(t)< (Gdata(i,"Fj")+1)),U(i,t))=e=0;
Dntime2(i)$(Gdata(i,"DT")>1)..
 sum(t$(ord(t)>24─Gdata(i,"DT")+1),1─U(i,t)─z(i,t))=g=0;
Dntime3(i,t)$(ord(t)>Gdata(i,"Fj") and ord(t)<24─Gdata(i,"DT")+2 and
not(Gdata(i,"Fj")>24─Gdata(i,"DT"))) .. sum(h$((ord(h)>ord(t)─1) and
(ord(h)<ord(t)+Gdata(i,"DT"))),1─U(i,h)) =g=Gdata(i,"DT")*z(i,t);
startc(i,t) .. StC(i,t)=g=Gdata(i,"costst")*y(i,t);
shtdnc(i,t) .. SDC(i,t)=g=Gdata(i,"CostsD")*z(i,t);
genconst1(i,h) .. p(i,h)=e=u(i,h)*Gdata(i,"Pmin")+sum(k,Pk(i,h,k));
Genconst2(i,h)$(ord(h)>0) .. U(i,h)=e=U(i,h─1)$(ord(h)>1)
                                   +Gdata(i,"Uini")$(ord(h)=1)+y(i,h)─z(i,h);
Genconst3(i,t,k) .. Pk(i,t,k)=l=U(i,t)*data(k,i,'DP');
Genconst4 .. costThermal=e=sum((i,t),StC(i,t)+SDC(i,t))
                                   +sum((t,i), u(i,t)*Gdata(i,'Mincost')
                                   +sum(k,data(k,i,'s')*pk(i,t,k)));
Ramp1(i,t) .. p(i,t─1)─p(i,t)=l=U(i,t)*Gdata(i,'RD')+z(i,t)*Gdata(i,"SD");
Ramp2(i,t) .. p(i,t)=l=pu(i,t);
Ramp3(i,t)$(ord(t)<24) .. pu(i,t)=l=(u(i,t)─z(i,t+1))*Gdata(i,"Pmax")
                                   +z(i,t+1)*Gdata(i,"SD");
Ramp4(i,t)$(ord(t)>1) .. pu(i,t)=l=p(i,t─1)+U(i,t─1)*Gdata(i,'RU')
                                   +y(i,t)*Gdata(i,"SU");
Balance(t)  ..  sum(i,p(i,t))=l= dataLP(t,'load');
Benefitcalc ..  OF=e=sum((i,t),dataLP(t,'lambda')*p(i,t))─ costThermal
Model UCLP / all /;
Option optcr=0.0;
Solve UCLP maximizing OF using mip ;
```

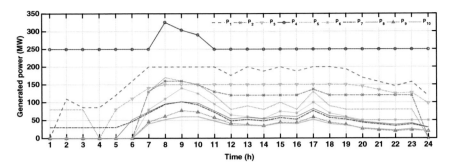

Fig. 5.7 Power schedules in price-based unit commitment vs time

The start-up, shut-down, and fuel costs are $499.000, $163.800, and $372,277.510, respectively. The net benefit of Genco is $58,186.325. The power schedules of thermal units in cost-based unit commitment are depicted in Fig. 5.7.

5.4   Applications

The unit commitment analysis has a vast range of applications, and some of them are described as follows:

5.4.1   Cost-Based UC

In practical applications, the thermal units might have some limitations in fuels. This means that the unit cannot generate more than some certain amount of energy. In other words, it might not be technically possible to keep the unit on for all time steps for the operating period. The fuel constrained UC is formulated in 4,335,178. Another important constraint in UC is environmental emission which is modeled in [4].

Another aspect of UC is considering the uncertainties associated with renewable energy resources (like wind turbines). Generally speaking, there are several methods for dealing with uncertainties in power system studies, such as

- Stochastic methods: two-stage scenario-based [5], two-point estimate [6, 12]
- Fuzzy methods [7]
- Robust optimization [8, 9]
- Information gap decision theory [10, 11]

The different uncertainty parameters and modeling methods are shown in Fig. 5.8 [12].

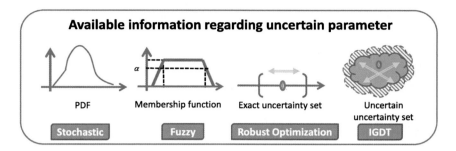

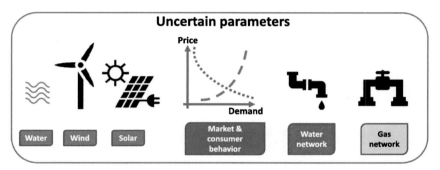

Fig. 5.8 Different uncertainty parameters and modeling methods

5.4.2   Price-Based UC

Some price-based UC formulations are listed as follows:

- Robust optimization-based self-scheduling of hydro-thermal Genco in smart grids [13]
- Risk averse optimal operation of a virtual power plant using two-stage stochastic programming [14]
- Smart self-scheduling of Gencos with thermal and energy storage units under price uncertainty [15].

References

1. A. Ademovic, S. Bisanovic, M. Hajro, A genetic algorithm solution to the unit commitment problem based on real-coded chromosomes and fuzzy optimization, in *Melecon 2010 - 2010 15th IEEE Mediterranean Electrotechnical Conference*, April 2010, pp. 1476–1481
2. J.M. Arroyo, A.J. Conejo, Optimal response of a thermal unit to an electricity spot market. IEEE Trans. Power Syst. **15**(3), 1098–1104 (2000)
3. M. Ortega-Vazquez, D. Kirschen, Optimizing the spinning reserve requirements using a cost/benefit analysis, in *2008 IEEE Power and Energy Society General Meeting - Conversion and Delivery of Electrical Energy in the 21st Century*, July 2008, p. 1

4. T. Gjengedal, Emission constrained unit-commitment (ECUC). IEEE Trans. Energy Convers. **11**(1), 132–138 (1996)
5. C. Uçkun, A. Botterud, J.R. Birge, An improved stochastic unit commitment formulation to accommodate wind uncertainty. IEEE Trans. Power Syst. **31**(4), 2507–2517 (2016)
6. A. Soroudi, M. Aien, M. Ehsan, A probabilistic modeling of photo voltaic modules and wind power generation impact on distribution networks. IEEE Syst. J. **6**(2), 254–259 (2012)
7. A. Soroudi, Possibilistic-scenario model for DG impact assessment on distribution networks in an uncertain environment. IEEE Trans. Power Syst. **27**(3), 1283–1293 (2012)
8. R. Jiang, J. Wang, Y. Guan, Robust unit commitment with wind power and pumped storage hydro. IEEE Trans. Power Syst. **27**(2), 800–810 (2012)
9. A. Soroudi, P. Siano, A. Keane, Optimal DR and ESS scheduling for distribution losses payments minimization under electricity price uncertainty. IEEE Trans. Smart Grid **7**(1), 261–272 (2016)
10. A. Soroudi, A. Rabiee, A. Keane, Information gap decision theory approach to deal with wind power uncertainty in unit commitment. Electr. Power Syst. Res. **145**, 137–148 (2017)
11. C. Murphy, A. Soroudi, A. Keane, Information gap decision theory-based congestion and voltage management in the presence of uncertain wind power. IEEE Trans. Sust. Energy **7**(2), 841–849 (2016)
12. A. Soroudi, T. Amraee, Decision making under uncertainty in energy systems: state of the art. Renew. Sust. Energ. Rev. **28**, 376–384 (2013)
13. A. Soroudi, Robust optimization based self scheduling of hydro-thermal Genco in smart grids. Energy **61**, 262–271 (2013)
14. M.A. Tajeddini, A. Rahimi-Kian, A. Soroudi, Risk averse optimal operation of a virtual power plant using two stage stochastic programming. Energy **73**, 958–967 (2014)
15. A. Soroudi, Smart self-scheduling of Gencos with thermal and energy storage units under price uncertainty. Int. Trans. Electr. Energy Syst. **24**(10), 1401–1418 (2014)

Chapter 6
Multi-Period Optimal Power Flow

This chapter provides a solution for optimal power flow OPF problem in GAMS. Different OPF models are investigated, such as single and multi-period DC-AC optimal power flow.

6.1 Single Period Optimal DC Power Flow

There are some necessary conditions that make the DC power flow acceptable as an approximate solution for AC power flow such as:

- The ratio of $\frac{x_{ij}}{r_{ij}}$ should be large enough that r_{ij} can be neglected.
- The voltage magnitudes are approximately 1 pu.

The DC power flow concept for a two-bus network is shown in Fig. 6.1. The basic variables in DC power flow are voltage angles δ_i. The angle of the slack bus is assumed to be zero as the reference for the rest of network.

The technical and economic characteristics of generating units shown in Table 6.1 are given as follows:

The demand at bus 2 is $L_2 = 400\,\text{MW}$, line reactance is $X_{12} = 0.2\,\text{pu}$, and line flow limit is $P_{12}^{\max} = 1.50\,\text{pu}$ (per unit on 100 MVA base). The optimization problem which should be solved is formulated in (6.1).

$$\min_{P_g, \delta_i} \text{OF} = \sum_{g_1, g_2} a_g (P_g)^2 + b_g P_g + c_g \tag{6.1a}$$

$$P_{ij} = \frac{\delta_1 - \delta_2}{X_{12}} \tag{6.1b}$$

$$P_{g_1} = P_{12} \tag{6.1c}$$

$$P_{g_2} + P_{12} = L_2 \tag{6.1d}$$

© Springer International Publishing AG 2017
A. Soroudi, *Power System Optimization Modeling in GAMS*,
DOI 10.1007/978-3-319-62350-4_6

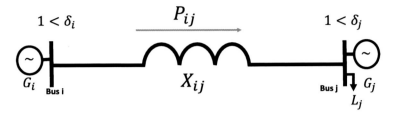

Fig. 6.1 DC power flow for a two-bus network

GCode 6.1 The OPF GAMS code for two-bus network, Example (6.1)

```
Sets
Gen /g1*g2/
bus /1*2/;
Scalars
L2      /400/
X12 /0.2/
Sbase /100/
P12_max /1.5/  ;
Table data(Gen,*)
        a      b      c      Pmin   Pmax
G1      3     20     100     28     206
G2    4.05  18.07  98.87    90     284;
Variables P(gen),OF,delta(bus),P12;
Equations
eq1,eq2,eq3,eq4;

eq1  ..  OF=e=sum(gen,data(gen,'a')*P(gen)*P(gen)+data(gen,'b')
*P(gen)+data(gen,'c'));
eq2  ..  P('G1')=e=P12;
eq3  ..  P('G2')+P12=e=L2/Sbase;
eq4  ..  P12=e=(delta('1')-delta('2'))/X12;
P.lo(gen)=data(gen,'Pmin')/Sbase;
P.up(gen)=data(gen,'Pmax')/Sbase;
P12.lo=-P12_max;
P12.up=+P12_max;
delta.fx('1')=0;
Model OPF /all/;
Solve OPF us qcp min of;
```

$$-P_{12}^{\max} \le P_{12} \le P_{12}^{\max} \qquad\qquad (6.1\text{e})$$

$$\delta_1 = 0 \quad \text{Slack} \qquad\qquad (6.1\text{f})$$

The GAMS code for solving the (6.1) is provided in GCode 6.1.

The operating costs would be \$306.108, $P_{12} = 150$ MW and $\delta_2 = -0.3$ (rad).

The general power flow concept is shown in Fig. 6.2. As it can be seen in Fig. 6.2, every bus might host some generation and demand. Each bus might be connected

Table 6.1 The techno-economic data of thermal units in two-bus OPF example

g	a_g (\$/MW^2)	b_g (\$/MW)	c_g (\$)	$P_g^{\min}$ (MW)	$P_g^{\max}$ (MW)
g_1	0.12	14.8	89	28	200
g_2	0.17	16.57	83	20	290

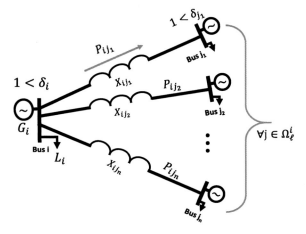

Fig. 6.2 General power flow concept

to other network buses by multiple branches (with different characteristics). The power balance between generation, demand and power transfers should be satisfied at every bus of the network under study.

The general mathematical formulation of OPF is described in (6.2). It should be noted that the technical terms of (6.2) are all linear. The only nonlinear part is the cost function. This makes the DC-OPF a strong tool for power system studies at the transmission level. This means that if the cost term can be expressed in linear form then the DC-OPF would become a linear programming problem and can be solved using linear solvers in GAMS such as CPLEX [1].

$$OF = \sum_{g \in \Omega_G} a_g(P_g)^2 + b_g P_g + c_g \tag{6.2a}$$

$$P_{ij} = \frac{\delta_i - \delta_j}{x_{ij}} \quad ij \in \Omega_\ell \tag{6.2b}$$

$$\sum_{g \in \Omega_G^i} P_g - L_i = \sum_{j \in \Omega_\ell^i} P_{ij} : \lambda_i \quad i \in \Omega_B \tag{6.2c}$$

$$-P_{ij}^{\max} \le P_{ij} \le P_{ij}^{\max} \quad ij \in \Omega_\ell \tag{6.2d}$$

$$P_g^{\min} \le P_g \le P_g^{\max} \tag{6.2e}$$

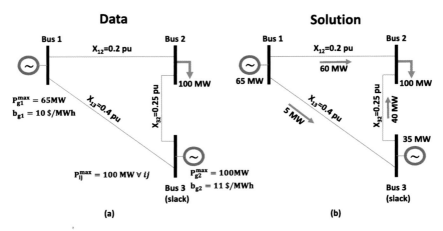

Fig. 6.3 Three bus network example (**a**) data and (**b**) power flow solution

6.1.1   Three-Bus Network DC-OPF

The three-bus network data is shown in Fig. 6.3. This example is taken from [2] (Example 4B, p. 110).

The GAMS code for solving this problem is given as GCode 6.2: The developed code for solving the OPF problem in three-bus network (Fig. 6.3) is explained here. It is general and can be used for any network with any size.

- Three sets are defined: bus (all network buses), slack(bus) which shows slack buses with reference angle values, Gen (set of generating units)
- A scalar value named *Sbase* is defined for per unit calculations
- The set node is defined as the similar set to set *bus*
- The table *GenData* defines the technical and economic characteristics of generating units
- The set *GBconect* defines the connection point of each generating unit
- The table *BusData* specifies the demand values in each bus
- The set *conex* specifies how each bus is connected to the other network buses
- The table *branch* defines the branch characteristics
- Four variables are used for this formulation namely: *OF* (objective function), *Pij* (active power flow between bus and node), *Pg*(generating schedule of each generating unit), and *delta* (voltage angle at each bus)
- Three equations are defined: *const1* (active flow calculation between each pair of connected buses), *const2* (nodal active power balance in each bus), and *const3* (objective function calculation)
- The definition of the model *loadflow* (specifies for GAMS that which constraints should be taken into account)
- Variables' limits specification

- Solve statement
- Generating report from the solved model
- The marginal value of *const2.m* provides the LMP at each bus. This is used for congestion cost calculation.

GCode 6.2 The DC-OPF GAMS code for three-bus network, Example (Sect. 6.1.1)

```
Sets
bus    /1*3/
slack(bus)  /3/
Gen /g1*g3/;
scalars
Sbase /100/;
alias(bus,node);
Table  GenData(Gen,*)    Generating  units  characteristics
     b     pmin pmax
g1  10     0     65
g2  11     0     100;
* ──────────────────────────────────────────────────
set GBconect(bus,Gen)  connectivity  index  of  each  generating  unit
    to  each  bus
/1      .     g1
 3      .     g2  /  ;
Table  BusData(bus,*) Demands  of  each  bus  in  MW
          Pd
2         100;
set conex               Bus  connectivity  matrix
/
1    .    2
2    .    3
1    .    3/;
conex(bus,node)$(conex(node,bus))=1;
Table  branch(bus,node,*)      Network  technical  characteristics
                 x            Limit
1    .    2     0.2          100
2    .    3     0.25         100
1    .    3     0.4          100  ;
branch(bus,node,'x')$(branch(bus,node,'x')=0)=branch(node,bus,'x'
    );
branch(bus,node,'Limit')$(branch(bus,node,'Limit')=0)=branch(node
    ,bus,'Limit');
branch(bus,node,'bij')$conex(bus,node)  =1/branch(bus,node,'x');
Variables
OF
Pij(bus,node)
Pg(Gen)
delta(bus);
Equations const1,const2,const3;
const1(bus,node)$conex(bus,node)..Pij(bus,node)=e=
                         branch(bus,node,'bij')*(delta(bus)−
                             delta(node));
```

```
const2(bus)   .. +sum(Gen$GBconect(bus,Gen),Pg(Gen))-BusData(bus,'
    pd')/Sbase=e=
                        +sum(node$conex(node,bus),Pij(bus,node)
                        );
const3        .. OF=g=sum(Gen,Pg(Gen)*GenData(Gen,'b')*Sbase);

Model loadflow        /const1,const2,const3/;
Pg.lo(Gen)=GenData(Gen,'Pmin')/Sbase;
Pg.up(Gen)=GenData(Gen,'Pmax')/Sbase;
delta.up(bus)=pi; delta.lo(bus)=-pi; delta.fx(slack)=0;
Pij.up(bus,node)$((conex(bus,node)))=1* branch(bus,node,'Limit')/
    Sbase;
Pij.lo(bus,node)$((conex(bus,node)))=-1*branch(bus,node,'Limit')/
    Sbase;
Solve loadflow minimizing OF using lp;
parameter report(bus,*),Congestioncost;
report(bus,'Gen(MW)')= sum(Gen$GBconect(bus,Gen),Pg.l(Gen))*sbase
    ;
report(bus,'Angle')=delta.l(bus);
report(bus,'load(MW)')= BusData(bus,'pd');
report(bus,'LMP($/MWh)')=const2.m(bus)/sbase  ;
Congestioncost= sum((bus,node),  Pij.l(bus,node)*(-const2.m(bus)+
    const2.m(node)))/2  ;
display report,Pij.l,Congestioncost;
```

The total operating costs will be 1035 \$/h and the LMP for all buses are equal to 11 \$/h. This means that if the demand value at any bus increases for 1 MW then the operating cost will increase by 11 \$/h. Since the generating unit 1 is generating power at its maximum limit (because it is cheaper) the additional demand should be supplied by generating unit 2 with operating cost equal to 11 \$/MW h. The detailed three-bus optimal power flow solution with ($P_{ij}^{\max} = 100$ MW) is given in Table 6.2.

Question: What would happen if the flow limit of the branch connecting bus 1 to bus 2 is reduced to 50 MW?

Answer: The answer is straightforward. The operating cost might increase but how much? We can easily decrease the flow limit in the GCode 6.2 (table Branch). The new operating cost would be 1056.250 \$/h. The LMP values are different for each bus in this case. The LMP values are $\lambda_1 = 10$ \$/MW h, $\lambda_2 = 11.625$ \$/MW h, $\lambda_3 = 11$ \$/MW h. The power flow solution is shown in Fig. 6.4. The LMP value in bus 1 is $\lambda_1 = 10$ \$/MW h because if the load increases in this node it will be supplied by generator 1 (which the operating costs are 10 \$/MW h). The LMP value in bus 3 is $\lambda_3 = 11$ \$/MW h because any increase in load in this node should be supplied by generator 2 (which the operating costs are 11 \$/MW h). Generator 1

Table 6.2 The three-bus optimal power flow solution ($P_{ij}^{\max} = 100$ MW)

Bus (i)	P_g (MW)	δ_i (rad)	L_i (MW)	λ_i (\$/MW h)
1	65	0.02	0	11
2	0	−0.10	100	11
3	35	0.00	0	11

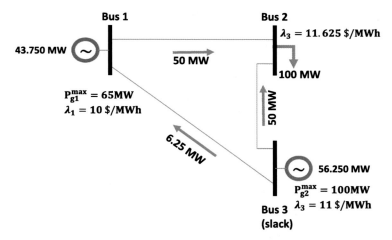

Fig. 6.4 The three-bus network with $P_{12}^{\max} = 50\,\text{MW}$

cannot send more power by line $1 - 2$ since it is congested and the direction of the power flow in branch $3 - 1$ is from bus 3 to bus 1. Finally, The LMP value in bus 2 is $\lambda_3 = 11.625\,\$/\text{MW h}$ because some of this demand will be supplied by generating unit 1 and some part will be supplied by unit 2. If the demand should pay the LMP value for every MW h then the total payment by the demand would be $100\,\text{MW} \times 11.625\,\$/\text{MW h} = 1162.5\,\$/\text{h}$. On the other hand, the generating units are also paid based on the LMP value of the connection point. The total payments to the generating units would be $43.75 \times 10 + 56.25 \times 11 = 1056.25\,\$/\text{h}$. As it can be seen, there is a difference between what demand pays and what generating units receive. This surplus money is equal to $1162.5 - 1056.25 = 106.25\,\$/\text{h}$. This is also called the congestion cost. Another technique for calculating the congestion costs (C_{cg}) is using the following formula [3]:

$$C_{cg} = \sum_{ij} P_{ji}(\lambda_i - \lambda_j) \tag{6.3}$$

The power flow solution of three-bus network with $P_{12}^{\max} = 50\,\text{MW}$ is shown in Fig. 6.4.

The detailed three-bus optimal power flow solution with ($P_{12}^{\max} = 50\,\text{MW}$) is given in Table 6.3.

Table 6.3 The three-bus
optimal power flow solution
($P_{12}^{max} = 50$ MW)

Bus (i)	P_g (MW)	δ_i (rad)	L_i (MW)	λ_i (\$/MW h)
1	43.75	−0.025	0	10
2	0	−0.125	100	11.625
3	56.25	0	0	11

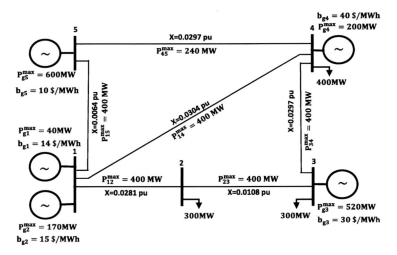

Fig. 6.5 The five-bus network data

6.1.2 Five-Bus Network DC-OPF

The five-bus network data is shown in Fig. 6.5. This example is taken from [4].

The GAMS code developed for solving OPF in five-bus network is provided in
GCode 6.3.

GCode 6.3 The OPF GAMS code for five-bus network, Example (Sect. 6.1.2)

```
Sets bus  /1*5/, slack(bus) /1/ , Gen /g1*g5/;
Scalars Sbase /100/ ; alias(bus,node);
Table GenData(Gen,*)
     b     pmin pmax
g1  14    0    40
g2  15    0    170
g3  30    0    520
g4  40    0    200
g5  20    0    600  ;
set GBconect(bus,Gen) connectivity index of generating unit
/1    .    g1
 1    .    g2
 3    .    g3
 4    .    g4
 5    .    g5 / ;
Table BusData(bus,*) Demands of each bus in MW
          Pd
```

```
2          300
3          300
4          400;
set conex            Bus  connectivity  matrix
/1    .    2
2     .    3
3     .    4
4     .    1
4     .    5
5     .    1/;
conex(bus,node)$conex(node,bus)=1;
Table  branch(bus,node,*)
                    x          Limit
1     .    2    0.0281      400
1     .    4    0.0304      400
1     .    5    0.0064      400
2     .    3    0.0108      400
3     .    4    0.0297      400
4     .    5    0.0297      240  ;
branch(bus,node,'x')$(branch(bus,node,'x')=0)=branch(node,bus,'x'
    );
branch(bus,node,'Limit')$(branch(bus,node,'Limit')=0)=branch(node
    ,bus,'Limit');
branch(bus,node,'bij')$conex(bus,node) =1/branch(bus,node,'x');
Variables OF, Pij(bus,node),Pg(Gen),delta(bus);
Equations const1,const2,const3;
const1(bus,node)$conex(bus,node)..  Pij(bus,node)=e=
 branch(bus,node,'bij')*(delta(bus)−delta(node));
const2(bus)  ..  +sum(Gen$GBconect(bus,Gen),Pg(Gen))−BusData(bus,'
    pd')/Sbase=e=
                              +sum(node$conex(node,bus),Pij(bus,
                                      node));
const3     ..  OF=g=sum(Gen,Pg(Gen)*GenData(Gen,'b')*Sbase);
Model  loadflow        /const1,const2,const3/;
Pg.lo(Gen)=GenData(Gen,'Pmin')/Sbase;
Pg.up(Gen)=GenData(Gen,'Pmax')/Sbase;
delta.up(bus)=pi;
delta.lo(bus)=−pi;
delta.fx(slack)=0;
Pij.up(bus,node)$((conex(bus,node)))= branch(bus,node,'Limit')/
    Sbase;
Pij.lo(bus,node)$((conex(bus,node)))=
−branch(bus,node,'Limit')/Sbase;
solve loadflow minimizing OF using lp;
parameter report(bus,*),Congestioncost;
report(bus,'Gen(MW)')= sum(Gen$GBconect(bus,Gen),Pg.l(Gen))*sbase
    ;
report(bus,'Angle')=delta.l(bus);
report(bus,'load(MW)')= BusData(bus,'pd');
report(bus,'LMP($/MWh)')=const2.m(bus)/sbase  ;
Congestioncost= sum((bus,node),
Pij.l(bus,node)*(−const2.m(bus)+const2.m(node)))/2;
Display report,Pij.l,Congestioncost;
```

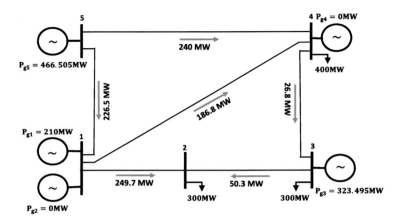

Fig. 6.6 The five-bus network flows and directions

Table 6.4 The optimal power flow solution in five-bus network

Bus (i)	P_g (MW)	δ_i (rad)	L_i (MW)	λ_i ($/MW h)
1	210.000	0.000	0	16.977
2	0.000	−0.070	300	26.384
3	323.495	−0.065	300	30.000
4	0.000	−0.057	400	39.943
5	466.505	0.014	0	10.000

The minimum operating cost is 17,479.897 $/h. The network is highly congested, and the LMP values are different in various buses. The five-bus network flows and directions are depicted in Fig. 6.6. The detailed five-bus optimal power flow solution is given in Table 6.4.

6.1.3 IEEE Reliability Test System 24 Bus

The IEEE RTS 24-bus network is shown in Fig. 6.7. It is a transmission network with the voltage levels of 138 kV, 230 kV, and Sbase = 100 MVA. The branch data for IEEE RTS 24-bus network is given in Table 6.5 [5]. The from bus, to bus, reactance (X), resistance (r), total line charging susceptance (b), and MVA rating (MVA) are specified in this table. The parallel lines in MATPOWER are merged, and the resultants are given in Table 6.5. The generation data for IEEE RTS 24-bus network is given in Table 6.6. The data of generating units in this network is inspired by Conejo et al. [6] and Bouffard et al. [7] with some modifications. The slack bus is bus 13 in this network.

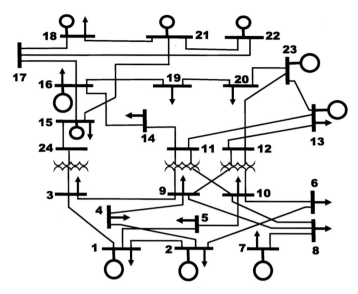

Fig. 6.7 IEEE RTS 24-bus network

Table 6.5 Branch data for IEEE RTS 24-bus network

From	To	r(pu)	x(pu)	b(pu)	Rating (MVA)	From	To	r(pu)	x(pu)	b(pu)	Rating (MVA)
1	2	0.0026	0.0139	0.4611	175	11	13	0.0061	0.0476	0.0999	500
1	3	0.0546	0.2112	0.0572	175	11	14	0.0054	0.0418	0.0879	500
1	5	0.0218	0.0845	0.0229	175	12	13	0.0061	0.0476	0.0999	500
2	4	0.0328	0.1267	0.0343	175	12	23	0.0124	0.0966	0.2030	500
2	6	0.0497	0.1920	0.0520	175	13	23	0.0111	0.0865	0.1818	500
3	9	0.0308	0.1190	0.0322	175	14	16	0.0050	0.0389	0.0818	500
3	24	0.0023	0.0839	0.0000	400	15	16	0.0022	0.0173	0.0364	500
4	9	0.0268	0.1037	0.0281	175	15	21	0.0032	0.0245	0.2060	1000
5	10	0.0228	0.0883	0.0239	175	15	24	0.0067	0.0519	0.1091	500
6	10	0.0139	0.0605	2.4590	175	16	17	0.0033	0.0259	0.0545	500
7	8	0.0159	0.0614	0.0166	175	16	19	0.0030	0.0231	0.0485	500
8	9	0.0427	0.1651	0.0447	175	17	18	0.0018	0.0144	0.0303	500
8	10	0.0427	0.1651	0.0447	175	17	22	0.0135	0.1053	0.2212	500
9	11	0.0023	0.0839	0.0000	400	18	21	0.0017	0.0130	0.1090	1000
9	12	0.0023	0.0839	0.0000	400	19	20	0.0026	0.0198	0.1666	1000
10	11	0.0023	0.0839	0.0000	400	20	23	0.0014	0.0108	0.0910	1000
10	12	0.0023	0.0839	0.0000	400	21	22	0.0087	0.0678	0.1424	500

Table 6.6 Generation data for IEEE RTS 24-bus network

Gen	Bus	P_i^{max}	P_i^{min}	b_i ($/MW)	Cs_i ($)	Cd_i ($)	RU_i (MW h^{-1})	RD_i (MW h^{-1})	SU_i (MW h^{-1})	SD_i (MW h^{-1})	UT_i(h)	DT_i(h)	$u_{i,t=0}$	U_i^0 (h)	S_i^0 (h)
g1	18	400	100	5.47	0	0	47	47	105	108	1	1	1	5	0
g2	21	400	100	5.47	0	0	47	47	106	112	1	1	1	6	0
g3	1	152	30.4	13.32	1430.4	1430.4	14	14	43	45	8	4	1	2	0
g4	2	152	30.4	13.32	1430.4	1430.4	14	14	44	57	8	4	1	2	0
g5	15	155	54.25	16	0	0	21	21	65	77	8	8	0	0	2
g6	16	155	54.25	10.52	312	312	21	21	66	73	8	8	1	10	0
g7	23	310	108.5	10.52	624	624	21	21	112	125	8	8	1	10	0
g8	23	350	140	10.89	2298	2298	28	28	154	162	8	8	1	5	0
g9	7	350	75	20.7	1725	1725	49	49	77	80	8	8	0	0	2
g10	13	591	206.85	20.93	3056.7	3056.7	21	21	213	228	12	10	0	0	8
g11	15	60	12	26.11	437	437	7	7	19	31	4	2	0	0	1
g12	22	300	0	0	0	0	35	35	315	326	0	0	1	2	0

6.1.3.1 IEEE-RTS: Base Case

In this case, it is assumed that the network is intact and all branches and generating units are working in normal condition. The minimum operating cost is \$29,574.275 obtained by using the GCode 6.4.

GCode 6.4 The OPF GAMS code for IEEE Reliability test 24-bus network, Example (Sect. 6.1.3)

```
sets bus /1*24/,slack(bus) /13/,Gen /g1*g12/; scalar Sbase
    /100/; alias(bus,node);
Table GenData(Gen,*)  Generating units characteristics
      Pmax Pmin    b      CostsD costst RU   RD   SU   SD   UT   DT
           uini U0   So
g1    400  100    5.47  0       0      47   47   105  108  1    1
      1    5      0    ;
set GBconect(bus,Gen) connectivity index of each generating unit
    to each bus
/18          .        g1 / ;
Table BusData(bus,*) Demands of each bus in MW/MVar
      Pd    Qd
1     108   22;
Table branch(bus,node,*)    Network technical characteristics
                 r      x       b      limit
1    .    2     0.0026  0.0139  0.4611 175;
parameter conex(bus,node);
conex(bus,node)$branch(bus,node,'limit')=1;
conex(bus,node)$(conex(node,bus))=1;
branch(bus,node,'x')$(branch(bus,node,'x')=0)=branch(node,bus,'x'
    );
branch(bus,node,'Limit')$(branch(bus,node,'Limit')=0)=branch(node
    ,bus,'Limit');
branch(bus,node,'bij')$conex(bus,node) =1/branch(bus,node,'x');
Variables OF,Pij(bus,node),Pg(Gen),delta(bus);
Equations const1,const2,const3;
const1(bus,node)$( conex(bus,node)) ..
Pij(bus,node)=e= branch(bus,node,'bij')*(delta(bus)-delta(node));
const2(bus) .. +sum(Gen$GBconect(bus,Gen),Pg(Gen))-BusData(bus,'
    pd')/Sbase=e=
+sum(node$conex(node,bus),Pij(bus,node));
const3      .. OF=g=sum(Gen,Pg(Gen)*GenData(Gen,'b')*Sbase);
Model loadflow      /const1,const2,const3/;
Pg.lo(Gen)=GenData(Gen,'Pmin')/Sbase;Pg.up(Gen)=GenData(Gen,
'Pmax')/Sbase;
delta.up(bus)=pi/2; delta.lo(bus)=-pi/2;delta.fx(slack)=0;
Pij.up(bus,node)$((conex(bus,node)))=1* branch(bus,node,'Limit')/
    Sbase;
Pij.lo(bus,node)$((conex(bus,node)))=-1*branch(bus,node,'Limit')/
    Sbase;
Solve loadflow minimizing OF using lp;
```

Table 6.7 Base case solution of IEEE RTS 24-bus network (branch flow limits are unchanged)

Bus	P_g (MW)	δ_i (rad)	Load (MW)	λ_i ($/MW h)
1	152	−0.150	108	20.7
2	152	−0.151	97	20.7
3		−0.113	180	20.7
4		−0.185	74	20.7
5		−0.191	71	20.7
6		−0.230	136	20.7
7	257.15	−0.105	125	20.7
8		−0.186	171	20.7
9		−0.136	175	20.7
10		−0.172	195	20.7
11		−0.044		20.7
12		−0.031		20.7
13	206.85	0.000	265	20.7
14		0.027	194	20.7
15	167	0.182	317	20.7
16	155	0.168	100	20.7
17		0.251		20.7
18	400	0.276	333	20.7
19		0.146	181	20.7
20		0.164	128	20.7
21	400	0.291		20.7
22	300	0.399		20.7
23	660	0.187		20.7
24		0.069		20.7

The base case solution of IEEE RTS 24-bus network without changing the branch flow limits is described in Table 6.7. As it can be observed in Table 6.7, the LMP values are all the same and equal to 20.7 $/MW h. This is because there is no congestion in this network for the given loading values. The congestion cost would be zero in this case.

6.1.3.2 IEEE-RTS: Branch Flow Limit Reduction

Now the branch flow limits are reduced by 30%. The problem is solved again, and the angle values are found as given in Table 6.8. The minimum operating cost is $29,747.745 obtained by using the GCode 6.4. The congestion cost is $4597.217 in this case. The solution of IEEE RTS 24-bus network (branch flow limits are reduced by 30%) is given in Table 6.8. Reducing the branch flow limit will not only increase the operating cost but also makes the LMP values different across the network.

Table 6.8 Solution of IEEE RTS 24-bus network (branch flow limits are reduced by 30%)

Bus	P_g (MW)	δ_i (rad)	Load (MW)	λ_i ($/MW h)
1	152.00	−0.162	108	20.66
2	152.00	−0.163	97	20.71
3		−0.129	180	19.11
4		−0.197	74	20.85
5		−0.203	71	20.98
6		−0.241	136	21.17
7	247.50	−0.130	125	20.70
8		−0.205	171	21.14
9		−0.147	175	20.96
10		−0.183	195	21.32
11		−0.053		22.69
12		−0.038		20.55
13	251.24	0.000	265	20.93
14		0.012	194	25.78
15	132.26	0.158	317	16.00
16	155.00	0.149	100	15.68
17		0.230		15.79
18	400.00	0.255	333	15.85
19		0.131	181	16.86
20		0.152	128	17.87
21	400.00	0.269		15.89
22	300.00	0.378		15.85
23	660.00	0.177		18.42
24		0.049		17.19

6.1.3.3 IEEE-RTS: Branch Outage

In this case, some branch contingencies are examined. In order to simulate the branch outage, the following equation should be satisfied.

$$P_{ij} - \frac{\delta_i - \delta_j}{x_{ij}} \leq M\xi_{ij} \tag{6.4}$$

$$P_{ij} - \frac{\delta_i - \delta_j}{x_{ij}} \geq -M\xi_{ij} \tag{6.5}$$

where ξ_{ij} is a binary parameter which shows the status of the branch connecting bus i to bus j. In the developed GAMS code, if the branch limit is set to 0 it is considered as an outaged branch. This is because $const1$ which models the flow calculation in line ij is calculated for every branch which has $conex(bus, node) > 0$. Now some different contingencies are evaluated as follows:

- Contingency 1: branch $\ell_{20-19}, \ell_{12-23}$ are out. Congestion costs are \$4905.000 and OF = \$29,888.196. The congested lines are ℓ_{13-23}. The LMP values are not the same in different buses.
- Contingency 2: branch $\ell_{14-16}, \ell_{16-19}$ are out. Congestion costs are \$6224.250 and OF = \$32,199.855. The congested lines are ℓ_{24-3}, ℓ_{8-7}. The LMP values are not the same in different buses.
- Contingency 3: branch ℓ_{1-5}, ℓ_{4-2} are out. Congestion costs are \$0 and OF = \$29,574.275. No line would be congested and therefore the LMP values are the same in all buses.

The OPF solutions for these three contingency cases are provided in Table 6.9.

Table 6.9 Solution of IEEE RTS 24-bus network (branch outage contingencies)

	Contingency 1			Contingency 2			Contingency 3		
Bus	P_g (MW)	δ_i (rad)	λ_i (\$/MW h)	P_g (MW)	δ_i (rad)	λ_i (\$/MW h)	P_g (MW)	δ_i (rad)	λ_i (\$/MW h)
1	152	−0.218	20.70	152	−0.106	20.93	152	−0.050	20.70
2	152	−0.219	20.70	152	−0.112	20.93	152	−0.054	20.70
3		−0.193	20.70		0.055	20.93		−0.088	20.70
4		− 0.252	20.70		− 0.156	20.93		−0.215	20.70
5		− 0.257	20.70		-0.173	20.93		−0.239	20.70
6		− 0.294	20.70		− 0.227	20.93		−0.210	20.70
7	289.15	−0.123	20.70	300	−0.038	20.70	257.15	−0.108	20.70
8		−0.224	20.70		− 0.145	20.93		−0.189	20.70
9		−0.202	20.70		−0.116	20.93		−0.138	20.70
10		−0.235	20.70		−0.181	20.93		−0.176	20.70
11		−0.094	20.70		−0.122	20.93		−0.044	20.70
12		−0.116	20.70		−0.037	20.93		−0.032	20.70
13	206.85		20.70	436		20.93	206.85		20.70
14		−0.053	20.70		−0.203	20.93		0.029	20.70
15	167	0.077	20.70	66.25	0.598	5.47	167	0.189	20.70
16	155	0.060	20.70	54.25	0.631	5.47	155	0.173	20.70
17		0.144	20.70		0.691	5.47		0.257	20.70
18	400	0.171	20.70	400	0.706	5.47	400	0.283	20.70
19		0.019	20.70		0.073	20.93		0.150	20.70
20		0.419	10.89		0.109	20.93		0.167	20.70
21	400	0.185	20.70	329.5	0.711	5.47	400	0.298	20.70
22	300	0.293	20.70	300	0.827	5.47	300	0.405	20.70
23	628	0.433	10.89	660	0.142	20.93	660	0.190	20.70
24		−0.026	20.70		0.390	5.47		0.083	20.70

6.1.3.4 IEEE-RTS: Generator Outage

In this case, some generating units are out of service. This can be because of unplanned outage or maintenance purpose. Now some different contingencies are evaluated as follows:

- Contingency 1: The generating unit g_9 (connected to bus 7) is out. Solving the problem shows that no line is congested but removing the g_9 will cause an increase in total operating costs which becomes OF = \$29,633.420.
- Contingency 2: The generating unit g_5 (connected to bus 15) is out. Solving the problem shows that line ℓ_{8-7} is congested but removing the g_5 will cause an increase in total operating costs which becomes OF = \$30,328.570. The LMP values are different on different buses, and the congestion costs are \$40.250.
- Contingency 3: The generating unit g_8 (connected to bus 23) is out. Solving the problem shows that line ℓ_{8-7} is congested but removing the g_8 will cause an increase in total operating costs which becomes OF = \$33,078.420. The LMP values are different on different buses, and the congestion costs are \$40.250.
- Contingency 4: The generating unit g_{10} (connected to bus 13) is out. The problem becomes infeasible, and GAMS cannot find any solution for it. This is because no solution is found which satisfies the technical constraints. Some load shedding action should take place.

The OPF solutions for first three contingency cases are given in Table 6.10. In order to analyze the contingency case 4, the load shedding is modeled using a virtual generating unit connected to bus i as shown in Fig. 6.8. The production cost of this specific unit is set to a high value. It is usually called the value of the loss of load. The production level of this generating unit is equal to the load shedding that should happen in bus i. The maximum power of this unit is equal to the load connected to the bus i ($0 \leq \mathrm{LS}_i \leq L_i$). The DC-OPF problem considering the load shedding concept is formulated in (6.6).

$$\mathrm{OF} = \sum_{g \in \Omega_G} a_g (P_g)^2 + b_g P_g + c_g + \sum_i \mathrm{VOLL} \times \mathrm{LS}_i \tag{6.6a}$$

$$P_{ij} = \frac{\delta_i - \delta_j}{x_{ij}} \quad ij \in \Omega_\ell \tag{6.6b}$$

$$\sum_{g \in \Omega_G^i} P_g + \mathrm{LS}_i - L_i = \sum_{j \in \Omega_\ell^i} P_{ij} : \lambda_i \quad i \in \Omega_B \tag{6.6c}$$

$$-P_{ij}^{\max} \leq P_{ij} \leq P_{ij}^{\max} \quad ij \in \Omega_\ell \tag{6.6d}$$

$$P_g^{\min} \leq P_g \leq P_g^{\max} \tag{6.6e}$$

$$0 \leq \mathrm{LS}_i \leq L_i \tag{6.6f}$$

Table 6.10 Solution of IEEE RTS 24-bus network (generator outage contingencies)

	Contingency 1			Contingency 2			Contingency 3		
	P_g			P_g			P_g		
Bus	(MW)	δ_i (rad)	λ_i ($/MW h)	(MW)	δ_i (rad)	λ_i ($/MW h)	(MW)	δ_i (rad)	λ_i ($/MW h)
1	152	−0.246	20.93	152	−0.171	20.93	152	−0.186	20.93
2	152	−0.248	20.93	152	−0.172	20.93	152	−0.187	20.93
3		−0.195	20.93		−0.156	20.93		−0.165	20.93
4		−0.283	20.93		−0.204	20.93		−0.219	20.93
5		−0.291	20.93		−0.208	20.93		−0.224	20.93
6		−0.331	20.93		−0.244	20.93		−0.261	20.93
7		−0.576	20.93	300	−0.058	20.7	300	−0.075	20.7
8		−0.499	20.93		−0.165	20.93		−0.182	20.93
9		−0.235	20.93		−0.153	20.93		−0.169	20.93
10		−0.275	20.93		−0.184	20.93		−0.202	20.93
11		−0.097	20.93		−0.065	20.93		−0.076	20.93
12		−0.082	20.93		−0.044	20.93		−0.071	20.93
13	464	0.000	20.93	319	0.000	20.93	514	0.000	20.93
14		−0.024	20.93		−0.020	20.93		−0.034	20.93
15	167	0.129	20.93	12	0.095	20.93	167	0.098	20.93
16	155	0.118	20.93	155	0.099	20.93	155	0.081	20.93
17		0.200	20.93		0.175	20.93		0.165	20.93
18	400	0.226	20.93	400	0.198	20.93	400	0.191	20.93
19		0.103	20.93		0.092	20.93		0.038	20.93
20		0.126	20.93		0.122	20.93		0.038	20.93
21	400	0.240	20.93	400	0.210	20.93	400	0.206	20.93
22	300	0.348	20.93	300	0.320	20.93	300	0.314	20.93
23	660	0.152	20.93	660	0.152	20.93	310	0.051	20.93
24		0.006	20.93		−0.001	20.93		−0.003	20.93

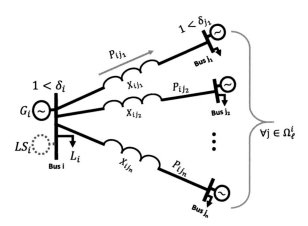

Fig. 6.8 Load shedding modeling using virtual generating unit

The GAMS code for modeling and calculating the load shedding values is given in GCode 6.5:

GCode 6.5 The OPF GAMS code for IEEE Reliability test 24-bus network with load shedding modeling

```
Sets
bus    /1*24/
slack(bus)  /13/
Gen  /g1*g12/
scalars
Sbase  /100/
VOLL  /10000/;
alias(bus,node);
table  GenData(Gen,*)    Generating  units  characteristics
       Pmax Pmin    b
********REMOVED FOR SAVING SPACE******
;
set  GBconect(bus,Gen)  connectivity  index  of  each  generating  unit
     to  each  bus
/
**REMOVED FOR SAVING SPACE******
***************************** /  ;
Table  BusData(bus,*)  Demands  of  each  bus  in  MW
       Pd    Qd
********REMOVED FOR SAVING SPACE******
;

table  branch(bus,node,*)      Network  technical  characteristics
                r      x       b        limit
********REMOVED FOR SAVING SPACE******
;
branch(bus,node,'x')$(branch(bus,node,'x')=0)=branch(node,bus,'x'
    );
branch(bus,node,'Limit')$(branch(bus,node,'Limit')=0)=branch(node
    ,bus,'Limit');
branch(bus,node,'bij')$branch(bus,node,'Limit')  =1/branch(bus,
    node,'x');
parameter  conex(bus,node);
conex(bus,node)$(branch(bus,node,'limit')and  branch(node,bus,'
    limit'))=1;
conex(bus,node)$(conex(node,bus))=1;
Variables
OF,  Pij(bus,node),Pg(Gen),delta(bus),lsh(bus);
Equations
const1,const2,const3;
const1(bus,node)$(  conex(bus,node))  ..  Pij(bus,node)=e=
                     branch(bus,node,'bij')*(delta(bus)−delta(node
                        ));
const2(bus)  ..  lsh(bus)  +sum(Gen$GBconect(bus,Gen),Pg(Gen))−
    BusData(bus,'pd')/Sbase
                                    =e=+sum(node$conex(node,bus
                                    ),Pij(bus,node));
```

```
const3      .. OF=g=sum((bus,Gen)$GBconect(bus,Gen),Pg(Gen)*GenData
    (Gen,'b')*Sbase)+sum(bus,VOLL*lsh(bus)*Sbase);
model loadflow        /const1,const2,const3/;
Pg.lo(Gen)=GenData(Gen,'Pmin')/Sbase;
Pg.up(Gen)=GenData(Gen,'Pmax')/Sbase;
delta.up(bus)=pi/2;
delta.lo(bus)=-pi/2;
delta.fx(slack)=0;
Pij.up(bus,node)$((conex(bus,node)))=1* branch(bus,node,'Limit')/
    Sbase;
Pij.lo(bus,node)$((conex(bus,node)))=-1*branch(bus,node,'Limit')/
    Sbase;
lsh.up(bus)= BusData(bus,'pd')/Sbase;
lsh.lo(bus)= 0;
solve loadflow minimizing OF using lp;
parameter report(bus,*),Congestioncost;
report(bus,'Gen(MW)')= 1*sum(Gen$GBconect(bus,Gen),Pg.l(Gen))*
    sbase;
report(bus,'Angle')=delta.l(bus);
report(bus,'load(MW)')= BusData(bus,'pd');
report(bus,'LSH')=lsh.l(bus)*sbase ;
report(bus,'LMP($/MWh)')=const2.m(bus)/sbase ;
Congestioncost= sum((bus,node), Pij.l(bus,node)*(-const2.m(bus)+
    const2.m(node)))/2 ;
display report,Pij.l,Congestioncost;
execute_unload "opf.gdx" report
execute 'gdxxrw.exe  OPF.gdx o=OPF.xls par=report rng=report!A1'
display GBconect;
```

The OPF solution for contingency case 4 is given in Table 6.11. The simulation results show that the demand in bus 18 should be reduced by 116 MW as given in Table 6.11. It should be noted that the LMP values are not meaningful in this case. The LMP values are calculated and shown in Table 6.12.

In order to calculate the LMP in a network with load shedding, the calculated amount of load shedding should be curtailed from the demand in the specific bus (here bus 18) and the problem should be resolved.

Nomenclature

Indices and Sets

g Index of thermal generating units.

i,j Index of network buses.

Ω_G Set of all thermal generating units.

Ω_G^i Set of all thermal generating units connected to bus i.

Ω_ℓ^i Set of all buses connected to bus i.

Ω_B Set of network buses.

Ω_ℓ Set of network branches.

Table 6.11 Solution of IEEE RTS 24-bus network (generator g_{10} outage with load shedding modeling)

Bus	P_g (MW)	δ_i (rad)	L_i (MW)	LSH_i (MW)
1	152	−0.095	108	
2	152	−0.097	97	
3		−0.042	180	
4		−0.132	74	
5		−0.140	71	
6		−0.181	136	
7	300	0.006	125	
8		−0.101	171	
9		−0.084	175	
10		−0.124	195	
11		−0.004		
12		−0.001		
13		0.000	265	
14		0.093	194	
15	215	0.285	317	
16	155	0.259	100	
17		0.360		
18	400	0.396	333	116
19		0.220	181	
20		0.222	128	
21	400	0.405		
22	300	0.511		
23	660	0.237		
24		0.160		

Parameters

L_i	Electric power demand in bus i
a_g, b_g, c_g	Fuel cost coefficients of thermal unit g.
$P_g^{max/min}$	Maximum/minimum limits of power generation of thermal unit g.
P_{ij}^{max}	Maximum power flow limits of branch connecting bus i to j.
x_{ij}	Reactance of branch connecting bus i to j.
r_{ij}	Resistance of branch connecting bus i to j.

Variables

P_{ij}	Active power flow of branch connecting bus i to j.
P_g	Active power generated by thermal unit g(MW).
λ_i	Locational marginal price in bus i ($/MW h).
LS_i	Load shedding in bus i (MW).
OF	Total operating costs ($/h).
δ_i	Voltage angle in bus i (rad).

Table 6.12 Solution of IEEE RTS 24-bus network (generator g_{10} outage with load shedding modeling and LMP calculation)

Bus	P_g (MW)	δ_i (rad)	L_i (MW)	λ_i ($/MW h)
1	152	−0.095	108	26.110
2	152	−0.097	97	26.110
3		−0.042	180	26.110
4		−0.132	74	26.110
5		−0.140	71	26.110
6		−0.181	136	26.110
7	300	0.006	125	20.700
8		−0.101	171	26.110
9		−0.084	175	26.110
10		−0.124	195	26.110
11		−0.004		26.110
12		−0.001		26.110
13		0.000	265	26.110
14		0.093	194	26.110
15	215	0.285	317	26.110
16	155	0.259	100	26.110
17		0.360		26.110
18	400	0.396	217	26.110
19		0.220	181	26.110
20		0.222	128	26.110
21	400	0.405		26.110
22	300	0.511		26.110
23	660	0.237		26.110
24		0.160		26.110

6.2   Multi-Period Optimal Wind-DC OPF

The load flow model considering wind power and load shedding is shown in Fig. 6.9. The multi-period wind DC-OPF is formulated as follows:

$$\text{OF} = \sum_{g,t} a_g (P_{g,t})^2 + b_g P_{g,t} + c_g + \sum_{i,t} \text{VOLL} \times \text{LS}_{i,t} + \text{VWC} \times P_{i,t}^{\text{wc}} \quad (6.7\text{a})$$

$$\sum_{g \in \Omega_G^i} P_{g,t} + \text{LS}_{i,t} + P_{i,t}^w - L_{i,t} = \sum_{j \in \Omega_\ell^i} P_{ij,t} : \lambda_{i,t} \quad (6.7\text{b})$$

$$P_{ij,t} = \frac{\delta_{i,t} - \delta_{j,t}}{x_{ij}} \quad (6.7\text{c})$$

$$-P_{ij}^{\max} \leq P_{ij,t} \leq P_{ij}^{\max} \quad (6.7\text{d})$$

$$P_g^{\min} \leq P_{g,t} \leq P_g^{\max} \quad (6.7\text{e})$$

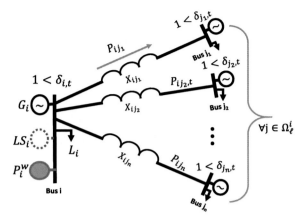

Fig. 6.9 Load flow model considering wind power and load shedding

$$P_{g,t} - P_{g,t-1} \leq \mathrm{RU}_g \tag{6.7f}$$

$$P_{g,t-1} - P_{g,t} \leq \mathrm{RD}_g \tag{6.7g}$$

$$0 \leq \mathrm{LS}_{i,t} \leq L_{i,t} \tag{6.7h}$$

$$P_{i,t}^{\mathrm{wc}} = w_t \Lambda_i^w - P_{i,t}^w \tag{6.7i}$$

$$0 \leq P_{i,t}^w \leq w_t \Lambda_i^w \tag{6.7j}$$

The objective function in (6.7a) consists of operating costs of thermal units, load shedding costs, and wind curtailment costs. The nodal power balance is formulated in (6.7b). This equation is also providing the LMP value in bus i at time t $(\lambda_{i,t})$. The active flow in branch connecting bus i to bus j is calculated in (6.7c). The branch flow limits of every branch is formulated in (6.7d). The operating limits of the thermal generating unit are modeled in (6.7e). The ramp rates of thermal units are described in (6.7f) and (6.7g). The load shedding of bus i is limited to the existing demand at time t as stated in (6.7h). The wind power curtailment is formulated in (6.7i). At any bus i hosting wind turbine, the amount of wind power generation depends on wind power availability and wind power capacity as described in (6.7j). It should bear in mind that the demand and wind power availability change vs time. These variations are reflected in $L_{i,t}$ and w_t, respectively. The wind power connection points to the network are indicated in Fig. 6.10.

The wind-demand variation pattern vs time is shown in Fig. 6.11. The optimal thermal unit power generation schedules of multi-period wind DC-OPF is given in Table 6.13.

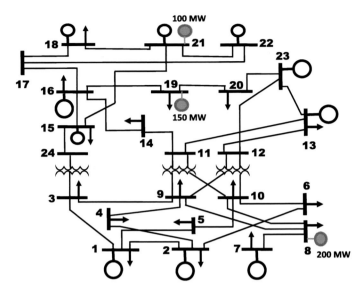

Fig. 6.10 Wind power connection points to the network

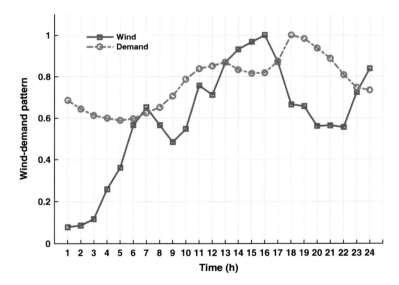

Fig. 6.11 Wind-demand variation patterns vs time

Table 6.13 Thermal unit power generation schedules of multi-period wind DC-OPF

Thermal unit	Bus	t_1	t_2	t_3	t_4	t_5	t_6	t_7	t_8	t_9	t_{10}	t_{11}	t_{12}	t_{13}	t_{14}	t_{15}	t_{16}	t_{17}	t_{18}	t_{19}	t_{20}	t_{21}	t_{22}	t_{23}	t_{24}
g1	18.0	4.0	4.0	3.5	3.1	2.7	2.3	2.6	3.1	3.5	4.0	4.0	4.0	4.0	3.6	3.1	3.1	3.5	4.0	4.0	4.0	4.0	3.5	3.1	2.6
g2	21.0	4.0	3.8	3.3	2.8	2.4	2.1	2.6	3.1	3.5	4.0	4.0	4.0	4.0	3.8	3.5	3.1	3.5	4.0	4.0	4.0	4.0	3.5	3.1	3.0
g3	1.0	0.4	0.3	0.3	0.3	0.3	0.3	0.3	0.3	0.3	0.4	0.4	0.4	0.4	0.3	0.3	0.4	0.6	0.7	0.9	0.7	0.6	0.4	0.3	0.3
g4	2.0	0.4	0.3	0.3	0.3	0.3	0.3	0.3	0.3	0.3	0.4	0.4	0.3	0.3	0.3	0.3	0.4	0.6	0.7	0.7	0.7	0.6	0.4	0.3	0.3
g5	15.0	0.5	0.5	0.5	0.5	0.5	0.5	0.5	0.5	0.5	0.6	0.5	0.5	0.5	0.5	0.5	0.5	0.8	1.0	1.0	1.2	1.0	0.8	0.5	0.5
g6	16.0	0.8	1.2	1.1	1.1	1.1	1.1	1.1	1.2	0.8	1.0	1.2	1.4	1.2	1.0	0.9	1.1	1.3	1.6	1.6	1.4	1.2	1.0	0.8	0.5
g7	23.0	1.4	1.2	1.1	1.1	1.1	1.1	1.1	1.2	1.4	1.6	1.8	2.0	2.2	2.1	2.3	2.5	2.7	2.9	3.1	2.9	2.7	2.5	2.3	2.1
g8	23.0	1.7	1.4	1.4	1.4	1.4	1.4	1.4	1.4	1.7	2.0	2.2	2.5	2.3	2.1	2.3	2.6	2.9	3.2	3.3	3.0	2.7	2.4	2.1	1.9
g9	7.0	0.8	0.8	0.8	0.8	0.8	0.8	0.8	0.8	0.8	0.8	0.8	0.8	0.8	0.8	0.8	0.8	1.2	1.7	1.2	1.1	0.9	0.8	0.8	0.8
g10	13.0	2.1	2.1	2.1	2.1	2.1	2.1	2.1	2.1	2.1	2.1	2.1	2.1	2.1	2.1	2.1	2.1	2.3	2.5	2.3	2.1	2.1	2.1	2.1	2.1
g11	15.0	0.1	0.1	0.1	0.1	0.1	0.1	0.1	0.1	0.1	0.1	0.1	0.1	0.1	0.1	0.1	0.1	0.2	0.3	0.2	0.1	0.1	0.1	0.1	0.1
g12	22.0	3.0	3.0	3.0	2.9	3.0	3.0	2.7	2.7	3.0	3.0	3.0	3.0	3.0	3.0	2.7	2.3	2.7	3.0	3.0	3.0	3.0	3.0	2.7	3.0

GCode 6.6 Multi-period DC-OPF for modified IEEE RTS 24-bus system

```
Sets  bus   /1*24/ , slack(bus) /13/ , Gen /g1*g12/ , t /t1*t24/;
Scalars Sbase /100/   ,VOLL /10000/ ,VWC /50/; alias(bus,node);
table GD(Gen,*)   Generating units characteristics
      Pmax Pmin    b        CostsD costst RU   RD   SU    SD    UT    DT
              uini U0    So
* Removed for saving space
Set GB(bus,Gen) connectivity index of each generating unit to
     each bus
/* Removed for saving space  / ;
Table BusData(bus,*) Demands of each bus in MW
*********Omitted for saving space   ;
table branch(bus,node,*)      Network technical characteristics
*********Omitted for saving space  ;
Table WD(t,*)
        w                 d
t1      0.0786666666666667 0.684511335492475
t2      0.0866666666666667 0.644122690036197
t3      0.117333333333333  0.61306915602972
t4      0.258666666666667  0.599733282530006
t5      0.361333333333333  0.588874071251667
t10     0.548              0.787007048961707
t20     0.561333333333333  0.936368832158506
t21     0.565333333333333  0.887597637645266
t22     0.556              0.809297008954087
t23     0.724              0.74585635359116
t24     0.84               0.733473042484283;
Parameter Wcap(bus)
/8      200
19      150
21      100/;
branch(bus,node,'x')$(branch(bus,node,'x')=0)=branch(node,bus,'x'
     );
branch(bus,node,'Limit')$(branch(bus,node,'Limit')=0)=branch(node
     ,bus,'Limit');
branch(bus,node,'bij')$branch(bus,node,'Limit') =1/branch(bus,
     node,'x');
branch(bus,node,'z')$branch(bus,node,'Limit')=sqrt(power(branch
(bus,node,'x'),2)+power(branch(bus,node,'r'),2));
branch(node,bus,'z')=branch(bus,node,'z');
parameter conex(bus,node);
conex(bus,node)$(branch(bus,node,'limit')and branch(node,bus,'
     limit'))=1;
conex(bus,node)$(conex(node,bus))=1;
Variables
OF, Pij(bus,node,t),Pg(Gen,t),delta(bus,t),lsh(bus,t),Pw(bus,t),pc
     (bus,t);
Equations
const1,const2,const3,const4,const5,const6;
const1(bus,node,t)$( conex(bus,node))  .. Pij(bus,node,t)=e=
branch(bus,node,'bij')*(delta(bus,t)−delta(node,t));
const2(bus,t)  .. lsh(bus,t)$BusData(bus,'pd')
+Pw(bus,t)$Wcap(bus)+sum(Gen$GB(bus,Gen),Pg(Gen,t))
```

```
−WD( t , ’d ’ )∗BusData ( bus , ’pd ’ ) / Sbase=e=+sum( node$conex ( node , bus ) ,
    Pij ( bus , node , t ) ) ;
const3       . .  OF=g=sum( ( bus , Gen , t )$GB( bus , Gen ) , Pg( Gen , t )∗GD( Gen , ’b
    ’ )∗Sbase )
                       +sum( ( bus , t ) , VOLL∗ lsh ( bus , t )∗ Sbase$BusData
                       +VWC∗Pc ( bus , t )∗sbase$Wcap( bus )   ) ;
const4 ( gen , t )   . .   pg( gen , t +1)−pg( gen , t )=1=GD( gen , ’RU’ ) / Sbase ;
const5 ( gen , t )   . .   pg( gen , t −1)−pg( gen , t )=1=GD( gen , ’RD’ ) / Sbase ;
const6 ( bus , t )$Wcap( bus )  . .   pc ( bus , t )=e=WD( t , ’w’ )∗Wcap( bus ) / Sbase−
    pw( bus , t ) ;
model  loadflow            / const1 , const2 , const3 , const4 , const5 , const6 / ;
Pg . lo ( Gen , t )=GD( Gen , ’Pmin ’ ) / Sbase ;  Pg . up( Gen , t )=GD( Gen , ’Pmax ’ ) /
    Sbase ;
delta . up( bus , t )=pi /2;  delta . lo ( bus , t )=−pi /2;  delta . fx ( slack , t )=0;
Pij . up( bus , node , t )$ ( ( conex ( bus , node ) ) )=1∗ branch ( bus , node , ’ Limit ’
    ) / Sbase ;
Pij . lo ( bus , node , t )$ ( ( conex ( bus , node ) ) )=−1∗branch ( bus , node , ’ Limit ’
    ) / Sbase ;
lsh . up( bus , t )=  WD( t , ’d ’ )∗BusData ( bus , ’pd ’ ) / Sbase ;  lsh . lo ( bus , t )=
    0;
Pw. up( bus , t )=WD( t , ’w’ )∗Wcap( bus ) / Sbase ;  Pw. lo ( bus , t )=0;
Pc . up( bus , t )=WD( t , ’w’ )∗Wcap( bus ) / Sbase ;  Pc . lo ( bus , t )=0;
Solve  loadflow  minimizing OF using lp ;
```

The wind power generations at efferent buses vs time are shown in Fig. 6.12.

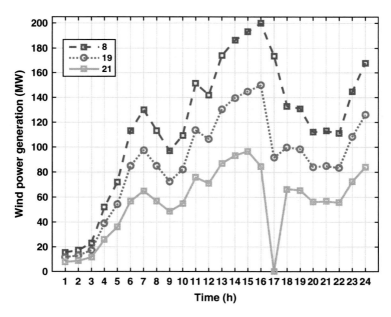

Fig. 6.12 Wind power generations at efferent buses vs time

The wind power curtailment is occurring in bus 19 at time t_{17} equal to 38.76 and also in bus 21 at time t_{16} equal to 15.32 MW and also at time t_{17} equal to 86.93 MW. The total operating cost (including the wind curtailment cost) would be \$4.3229 × 10^5.

Nomenclature

Indices and Sets

g	Index of thermal generating units.
w	Index of wind turbine units.
i, j	Index of network buses.
Ω_G	Set of all thermal generating units.
Ω_G^i	Set of all thermal generating units connected to bus i.
Ω_ℓ^i	Set of all buses connected to bus i.
Ω_B	Set of network buses.
Ω_ℓ	Set of network branches.

Parameters

$w_{i,t}$	Availability of wind turbine connected to bus i at time t
Λ_i^w	Capacity of wind turbine connected to bus i
$L_{i,t}$	Electric power demand in bus i at time t
a_g, b_g, c_g	Fuel cost coefficients of thermal unit g.
$P_g^{\max/\min}$	Maximum/minimum limits of power generation of thermal unit g.
$P_{ij}^{\max}$	Maximum power flow limits of branch connecting bus i to j.
x_{ij}	Reactance of branch connecting bus i to j.
r_{ij}	Resistance of branch connecting bus i to j.
VOLL	Value of loss of load (\$/MW h)
VWC	Value of loss of wind (\$/MW h)

Variables

$P_{ij,t}$	Active power flow of branch connecting bus i to j at time t (MW).
$P_{g,t}$	Active power generated by thermal unit g at time t (MW) .
$P_{i,t}^w$	Active power generated by wind turbine connected to bus i at time t (MW).
$P_{i,t}^{wc}$	Curtailed power of wind turbine connected to bus i at time t (MW).
$\lambda_{i,t}$	Locational marginal price in bus i at time t (\$/MW h).
$LS_{i,t}$	Load shedding in bus i at time t (MW).
OF	Total operating costs (\$).
$\delta_{i,t}$	Voltage angle of bus i at time t (rad).

6.3 Multi-Period Optimal AC Power Flow

The AC power flow considering wind generation and load shedding is shown in Fig. 6.13.

The optimal power flow equations are described in (6.8) as follows:

$$OF = \sum_{i,t} a_g (P_{i,t}^g)^2 + b_g P_{i,t}^g + c_g + \sum_{i,t} VOLL \times P_{i,t}^{LS} + VWC \times P_{i,t}^{wc} \tag{6.8a}$$

$$P_{i,t}^g + P_{i,t}^{LS} + P_{i,t}^w - P_{i,t}^L = \sum_{j \in \Omega_\ell^i} P_{ij,t} : \lambda_{i,t}^p \tag{6.8b}$$

$$Q_{i,t}^g + Q_{i,t}^{LS} + Q_{i,t}^w - Q_{i,t}^L = \sum_{j \in \Omega_\ell^i} Q_{ij,t} : \lambda_{i,t}^q \tag{6.8c}$$

$$I_{ij,t} = \frac{V_{i,t} \angle \delta_{i,t} - V_{j,t} \angle \delta_{j,t}}{Z_{ij} \angle \theta_{ij}} + \frac{bV_{i,t}}{2} \angle \left(\delta_{i,t} + \frac{\pi}{2} \right) \tag{6.8d}$$

$$S_{ij,t} = (V_{i,t} \angle \delta_{i,t}) I_{ij,t}^* \tag{6.8e}$$

$$P_{ij,t} = real\{S_{ij,t}\} = \frac{V_{i,t}^2}{Z_{ij}} \cos(\theta_{ij}) - \frac{V_{i,t}V_{j,t}}{Z_{ij}} \cos(\delta_{i,t} - \delta_{j,t} + \theta_{ij}) \tag{6.8f}$$

$$Q_{ij,t} = Img\{S_{ij,t}\} = \frac{V_{i,t}^2}{Z_{ij}} \sin(\theta_{ij}) - \frac{V_{i,t}V_{j,t}}{Z_{ij}} \sin(\delta_{i,t} - \delta_{j,t} + \theta_{ij}) - \frac{bV_{i,t}^2}{2} \tag{6.8g}$$

$$-S_{ij}^{max} \leq S_{ij,t} \leq S_{ij}^{max} \tag{6.8h}$$

$$P_i^{g,min} \leq P_{i,t}^g \leq P_i^{g,max} \tag{6.8i}$$

$$Q_i^{g,min} \leq Q_{i,t}^g \leq Q_i^{g,max} \tag{6.8j}$$

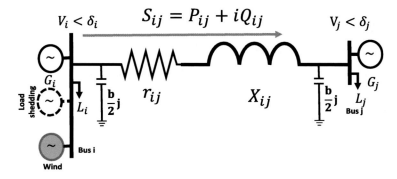

Fig. 6.13 AC power flow considering wind generation and load shedding

$$P_{i,t}^g - P_{i,t-1}^g \le \mathrm{RU}_g \tag{6.8k}$$

$$P_{i,t-1}^g - P_{i,t}^g \le \mathrm{RD}_g \tag{6.8l}$$

$$0 \le P_{i,t}^{\mathrm{LS}} \le P_{i,t}^L \tag{6.8m}$$

$$0 \le Q_{i,t}^{\mathrm{LS}} \le Q_{i,t}^L \tag{6.8n}$$

$$P_{i,t}^{\mathrm{wc}} = w_t \Lambda_i^w - P_{i,t}^w \tag{6.8o}$$

$$0 \le P_{i,t}^w \le w_t \Lambda_i^w \tag{6.8p}$$

The GAMS code developed for solving the minimum cost OPF is given in GCode 6.7. The optimal active/reactive power generation of thermal units in MP-AC OPF are given in Figs. 6.14 and 6.15, respectively.

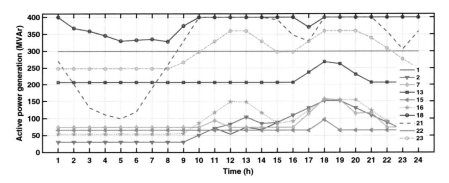

Fig. 6.14 Active power generation of thermal units in MP-AC OPF

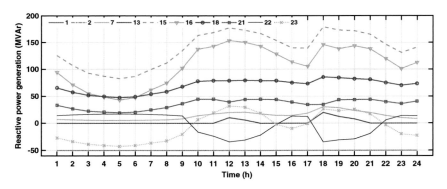

Fig. 6.15 Reactive power generation of thermal units in MP-AC OPF

GCode 6.7 Multi-period AC OPF for modified IEEE RTS 24 bus system

```
Sets  i    network  buses  /1*24/ ,slack(i)  /13/,  t  /t1*t24/
GB(i)  generating  buses  /1,2,7,15*16,18,21*23/;
scalars  Sbase  /100/  ,VOLL  /10000/,VWC  /50/;  alias(i,j);
Table  GenD(i,*)    Generating  units  characteristics
       pmax  pmin    b       Qmax  Qmin  Vg     RU    RD
1      152   30.4    13.32   192   −50   1.035  21    21
2      152   30.4    13.32   192   −50   1.035  21    21
7      350   75      20.7    300   0     1.025  43    43
13     591   206.85  20.93   591   0     1.02   31    31
15     215   66.25   21      215   −100  1.014  31    31
16     155   54.25   10.52   155   −50   1.017  31    31
18     400   100     5.47    400   −50   1.05   70    70
21     400   100     5.47    400   −50   1.05   70    70
22     300   0       0       300   −60   1.05   53    53
23     360   248.5   10.52   310   −125  1.05   31    31          ;
Table  BD(i,*)  Demands  of  each  bus  in  MW
       Pd    Qd
*  removed  for  saving  space;
table  LN(i,j,*)       Network  technical  characteristics
                  r       x        b          limit
*  removed  for  saving  space;
Table  WD(t,*)
           w                         d
*  removed  for  saving  space  ;
parameter  Wcap(i)
/8     200
19     150
21     100/;
LN(i,j,'x')$(LN(i,j,'x')=0)=LN(j,i,'x');LN(i,j,'r')$(LN(i,j,'r')
    =0)=LN(j,i,'r');
LN(i,j,'b')$(LN(i,j,'b')=0)=LN(j,i,'b');LN(i,j,'bij')$LN(i,j,'
    Limit')=1/LN(i,j,'x');
LN(i,j,'Limit')$(LN(i,j,'Limit')=0)=LN(j,i,'Limit');
LN(i,j,'z')$LN(i,j,'Limit')=sqrt(power(LN(i,j,'x'),2)+power(LN(i,
    j,'r'),2));
LN(j,i,'z')$(LN(j,i,'z')=0)=LN(i,j,'z');
LN(i,j,'th')$(LN(i,j,'Limit')  and  LN(i,j,'x')  and  LN(i,j,'r'))
                              =arctan(LN(i,j,'x')/(LN(i,j,'r')))
                                     ;
LN(i,j,'th')$(LN(i,j,'Limit')  and  LN(i,j,'x')  and  LN(i,j,'r')=0)
    =pi/2;
LN(i,j,'th')$(LN(i,j,'Limit')  and  LN(i,j,'r')  and  LN(i,j,'x')=0)
    =0;
LN(j,i,'th')$LN(i,j,'Limit')=LN(i,j,'th');  Parameter  cx(i,j);
cx(i,j)$(LN(i,j,'limit')and  LN(j,i,'limit'))=1;  cx(i,j)$(cx(j,i))
    =1;
Variables  OF,Pij(i,j,t),Qij(i,j,t),Pg(i,t),Qg(i,t),Va(i,t),V(i,t)
    ,Pw(i,t);
Equations  eq1,eq2,eq3,eq4,eq5,eq6,eq7;
eq1(i,j,t)$cx(i,j)..Pij(i,j,t)=e=(V(i,t)*V(i,t)*cos(LN(j,i,'th'))
                −V(i,t)*V(j,t)*cos(Va(i,t)−Va(j,t)+LN(j,i,'th'))
                )/LN(j,i,'z');
```

```
eq2(i,j,t)$cx(i,j).. Qij(i,j,t)=e=(V(i,t)*V(i,t)*sin(LN(j,i,'th'))
                       -V(i,t)*V(j,t)*sin(Va(i,t)-Va(j,t)+LN(j,i,'th'))
                       )/LN(j,i,'z')
                                                          -LN(j,i,'b')*V(
                                                          i,t)*V(i,t)
                                                          /2;
eq3(i,t)  .. Pw(i,t)$Wcap(i)+Pg(i,t)$GenD(i,'Pmax')-WD(t,'d')*BD(i
   ,'pd')/Sbase=e=
+sum(j$cx(j,i),Pij(i,j,t)));
eq4(i,t)  ..                  Qg(i,t)$GenD(i,'Qmax')-WD(t,'d')*BD(i,'
   qd')/Sbase=e=
+sum(j$cx(j,i),Qij(i,j,t)));
eq5      ..  OF=g=sum((i,t),Pg(i,t)*GenD(i,'b')*Sbase$GenD(i,'
   Pmax'));
eq6(i,t)$(GenD(i,'Pmax') and ord(t)>1) ..Pg(i,t)-Pg(i,t-1)=l=GenD
   (i,'RU')/Sbase;
eq7(i,t)$(GenD(i,'Pmax') and ord(t)<card(t)) ..Pg(i,t)-Pg(i,t+1)=
   l=
                                     GenD(i,'RD')/Sbase;
Model loadflow    /eq1,eq2,eq3,eq4,eq5,eq6,eq7/;
Pg.lo(i,t)=GenD(i,'Pmin')/Sbase; Pg.up(i,t)=GenD(i,'Pmax')/Sbase;
Qg.lo(i,t)=GenD(i,'Qmin')/Sbase; Qg.up(i,t)=GenD(i,'Qmax')/Sbase;
Va.up(i,t)=pi/2; Va.lo(i,t)=-pi/2; Va.l(i,t)=0; Va.fx(slack,t)=0;
Pij.up(i,j,t)$((cx(i,j)))=+1*LN(i,j,'Limit')/Sbase;
Pij.lo(i,j,t)$((cx(i,j)))=-1*LN(i,j,'Limit')/Sbase;
Qij.up(i,j,t)$((cx(i,j)))=+1*LN(i,j,'Limit')/Sbase;
Qij.lo(i,j,t)$((cx(i,j)))=-1*LN(i,j,'Limit')/Sbase; V.lo(i,t)
   =0.9; V.up(i,t)=1.1;
Pw.up(i,t)=WD(t,'d')*Wcap(i)/sbase; Pw.lo(i,t)=0;
Solve loadflow minimizing OF using nlp;
```

Nomenclature

Parameters

$w_{i,t}$	Availability of wind turbine connected to bus i at time t
Λ_i^w	Capacity of wind turbine connected to bus i
$P_{i,t}^L$	Active power component of demand in bus i at time t
$P_{i,t}^w$	Active power generation by wind turbine connected to bus i at time t
$Q_{i,t}^w$	Reactive power generation by wind turbine connected to bus i at time t
a_g, b_g, c_g	Fuel cost coefficients of thermal unit g.
$P_i^{g,max/min}$	Maximum/minimum limits of power generation of thermal unit g connected to bus i.
P_{ij}^{max}	Maximum power flow limits of branch connecting bus i to j.
x_{ij}	Reactance of branch connecting bus i to j.
r_{ij}	Resistance of branch connecting bus i to j.
VOLL	Value of loss of load (\$/MW h)

VWC — Value of loss of wind ($/MW h)

b — Total line charging susceptance of branch connecting bus i to j (pu).

Variables

$S_{ij,t}$ — Apparent power flow of branch connecting bus i to j at time t.

$P_{i,t}^{g}$ — Active power generated by thermal unit g connected to bus i at time t (MW).

$P_{i,t}^{w}$ — Active power generated by wind turbine connected to bus i at time t (MW).

$\lambda_{i,t}^{p}$ — Active locational marginal price (LMP) in bus i at time t ($/MW h).

$P_{i,t}^{LS}$ — Active Load shedding in bus i at time t (MW).

$P_{i,t}^{wc}$ — Curtailed power of wind turbine connected to bus i at time t (MW).

$I_{ij,t}$ — Current flow of branch connecting bus i to j at time t.

$Q_{i,t}^{LS}$ — Reactive Load shedding in bus i at time t (MVAr).

$Q_{i,t}^{g}$ — Reactive power generated by thermal unit g connected to bus i at time t (MW).

$\lambda_{i,t}^{q}$ — Reactive locational marginal price (LMP) in bus i at time t ($/MVArh).

OF — Total operating costs ($).

$\delta_{i,t}$ — Voltage angle in bus i at time t (rad).

$V_{i,t}$ — Voltage magnitude in bus i at time t (pu).

References

1. A. Meeraus A. Brooke, D. Kendrick, R. Raman, *GAMS/Cplex 7.0 User Notes*. GAMS Development Corp. (2000)
2. A.J. Wood, B.F. Wollenberg, *Power Generation, Operation, and Control* (Wiley, Hoboken, 2012)
3. P. Maghouli, A. Soroudi, A. Keane, Robust computational framework for mid-term techno-economical assessment of energy storage. IET Gener. Transm. Distrib. **10**(3), 822–831 (2016)
4. F. Li, R. Bo Small test systems for power system economic studies, in: *IEEE PES General Meeting* (2010), pp. 1–4
5. R.D. Zimmerman, C.E. Murillo-Sanchez, R.J. Thomas, Matpower: steady-state operations, planning, and analysis tools for power systems research and education. IEEE Trans. Power Syst. **26**(1), 12–19 (2011)
6. A.J. Conejo, M. Carrión, J.M. Morales, *Decision Making Under Uncertainty in Electricity Markets*, vol. 1 (Springer, New York, 2010)
7. F. Bouffard, F.D. Galiana, A.J. Conejo, Market-clearing with stochastic security-part II: case studies. IEEE Trans. Power Syst. **20**(4), 1827–1835 (2005)

Chapter 7
Energy Storage Systems

This chapter provides a solution for operation and planning aspects of energy storage systems (ESS) problem in GAMS. The ESS integration has been analyzed in operation and planning horizon. The inputs are generator's characteristics, electricity prices, demands, and network topology. The outputs of this code are operating schedules of ESS as well as the investment decisions.

7.1 Introduction

The energy storage system (ESS) is an attractive option to increase the flexibility of energy system operation and planning. These units can absorb energy in case of low electricity price or excessive generation and return it back in high price/low generation periods. Different technologies and objective functions have been reported in the literature.

Some of the ESS technologies are as follows:

- Superconducting magnetic energy storage system (SMES) [1]
- Compressed Air Energy Storage (CAES) [2]
- Super-capacitor Energy Storage [3]
- Pumped Hydro Storage [4]
- Battery energy storage [5]
- Flywheel Energy Storage System [6]
- Power to gas storage method [7]

Some of the objective functions in ESS studies are listed as follows:

- Compensate grid voltage fluctuations [1]
- Overcome the Destabilizing Effect of Instantaneous Constant Power Loads in DC Microgrids [3]
- Prevention of Transient Under-Frequency Load Shedding [8]

© Springer International Publishing AG 2017
A. Soroudi, *Power System Optimization Modeling in GAMS*,
DOI 10.1007/978-3-319-62350-4_7

- Reliability enhancement [9]
- Wind uncertainty management [10]
- Fault Ride Through Support of Grid-Connected VSC HVDC-Based Offshore Wind Farms [6]
- Phase balancing [11]
- Wind curtailment reduction and congestion management [12]
- Active power loss payment minimization [13]

7.2 ESS Operation

7.2.1 ESS Operation in DED

The total operating costs without using ESS are TC $= \$6.4796 \times 10^5$ as obtained in GCode 4.1. The DED problem integrated with ESS is formulated in (7.1).

$$\min_{P_{g,t}, \text{SOC}_t, P_t^d, P_t^c} \text{TC} = \sum_{g,t} a_g P_{g,t}^2 + b_g P_{g,t} + c_g \tag{7.1a}$$

$$P_g^{\min} \leq P_{g,t} \leq P_g^{\max} \tag{7.1b}$$

$$P_{g,t} - P_{g,t-1} \leq \text{RU}_g \tag{7.1c}$$

$$P_{g,t-1} - P_{g,t} \leq \text{RD}_g \tag{7.1d}$$

$$\text{SOC}_t = \text{SOC}_{t-1} + \left(P_t^c \eta_c - P_t^d / \eta_d \right) \Delta_t \tag{7.1e}$$

$$P_{\min}^c \leq P_t^c \leq P_{\max}^c \tag{7.1f}$$

$$P_{\min}^d \leq P_t^d \leq P_{\max}^d \tag{7.1g}$$

$$\text{SOC}_{\min} \leq \text{SOC}_t \leq \text{SOC}_{\max} \tag{7.1h}$$

$$\sum_g P_{g,t} + P_t^d \geq L_t - P_t^c \tag{7.1i}$$

The simulation parameters for this problem are given in Table 7.1. The economic dispatch data for four units example is given in Table 7.2. This table has 11 columns.

Table 7.1 The data for ESS integrated dynamic economic dispatch

Parameter	Value
SOC_0	100 MW
$\text{SOC}_{\max}$	300 MW
$P_{\max}^d$	0.2 $\text{SOC}_{\max}$
$P_{\min}^d$	0
$P_{\max}^c$	0.2 $\text{SOC}_{\max}$
$P_{\min}^c$	0
η_c	95%
η_d	90%

Table 7.2 The dynamic economic dispatch data for four thermal units example

Unit	a_g ($/MW^2)	b_g ($/MW)	c_g ($)	d_g (kg/MW^2)	e_g (kg/MW)	f_g (kg)	P_g^{min} (MW)	P_g^{max} (MW)	RU_g (MW)	RD_g (MW)
$g1$	0.12	14.8	89	1.2	−5	3	28	200	40	40
$g2$	0.17	16.57	83	2.3	−4.24	6.09	20	290	30	30
$g3$	0.15	15.55	100	1.1	−2.15	5.69	30	190	30	30
$g4$	0.19	16.21	70	1.1	−3.99	6.2	20	260	50	50

The first column is showing the generating unit index. The next three columns indicate the cost coefficients for these thermal units (a_g, b_g, c_g). The next three columns indicate the emission coefficients for these thermal units (d_g, e_g, f_g). The next two columns give the minimum and maximum generating limits of each unit if they ton. The last two columns indicate the ramp up/down rates of thermal units.

The GAMS code for solving the example (7.1) is given in Code 7.1.

GCode 7.1 The cost-based dynamic economic dispatch integrated with ESS (7.1)

```
Set       t            hours              /t1*t24/
Set       g            thermal units      /p1*p4/;

Table  gendata(g,*)  generator  cost  characteristics  and  limits
        a      b        c     d     e     f     Pmin  Pmax   RU0    RD0
p1     0.12  14.80   89    1.2  −5     3     28    200    40     40
p2     0.17  16.57   83    2.3  −4.24  6.09  20    290    30     30
p3     0.15  15.55  100    1.1  −2.15  5.69  30    190    30     30
p4     0.19  16.21   70    1.1  −3.99  6.2   20    260    50     50;

Table  data(t,*)
        lambda  load  wind
t1      32.71   510   44.1
t2      34.72   530   48.5
t3      32.71   516   65.7
t4      32.74   510   144.9
t5      32.96   515   202.3
t6      34.93   544   317.3
t7      44.9    646   364.4
t8      52      686   317.3
t9      53.03   741   271
t10     47.26   734   306.9
t11     44.07   748   424.1
t12     38.63   760   398
t13     39.91   754   487.6
t14     39.45   700   521.9
t15     41.14   686   541.3
t16     39.23   720   560
t17     52.12   714   486.8
t18     40.85   761   372.6
t19     41.2    727   367.4
t20     41.15   714   314.3
t21     45.76   618   316.6
t22     45.59   584   311.4
t23     45.56   578   405.4
t24     34.72   544   470.4;

Variables          OBJ               Objective (revenue)
                   costThermal       Cost of thermal units
                   p(g,t)            Power generated by thermal power
                                     plant
                   SOC(t) ,Pd(t),Pc(t)                              ;
p.up(g,t) = gendata(g,"Pmax") ;
p.lo(g,t) = gendata(g,"Pmin");
scalar SOC0 /100/, SOCmax /300/, eta_c /0.95/, eta_d /0.9/;
```

```
SOC.up(t)=SOCmax; SOC.lo(t)=0.2*SOCmax; SOC.fx('t24')=SOC0;
Pc.up(t)=0.2*SOCmax; Pc.lo(t)=0; Pd.up(t)=0.2*SOCmax; Pd.lo(t)=0;
Equations Genconst3,Genconst4,costThermalcalc,balance, constESS1;
costThermalcalc.. costThermal =e=sum((t,g), gendata(g,'a')*power
(p(g,t),2)
+gendata(g,'b')*p(g,t) +gendata(g,'c'));
Genconst3(g,t)  .. p(g,t+1)-p(g,t)=l=gendata(g,'RU0');
Genconst4(g,t)  .. p(g,t-1)-p(g,t)=l=gendata(g,'RD0');
constESS1(t)    .. SOC(t)=e=SOC0$(ord(t)=1)+ SOC(t-1)$(ord(t)>1)
                                           +Pc(t)*eta_c-Pd(t)/
                                           eta_d;
balance(t)  .. sum(g,p(g,t))+Pd(t)=g=data(t,'load')+Pc(t);

Model DEDESScostbased /all/;

Solve DEDESScostbased us qcp min costThermal;
```

The total operating costs with using ESS are TC = \$6.4553 × 10^5 as obtained in
GCode 7.1. The optimal solution for ESS integrated dynamic economic dispatch is
given in Table 7.3. The hourly dispatch of ESS in DED problem is shown in Fig. 7.1.
The hourly schedules of thermal units in DED-ESS problem are shown in Fig. 7.2.

Table 7.3 The ESS integrated dynamic economic dispatch

Time	L_t (MW)	$P_{g,t}$ (MW)	SOC_t (MW h)	P_t^d (MW)	P_t^c (MW)
t_1	510.00	555.92	143.62	0.00	45.92
t_2	530.00	555.92	168.25	0.00	25.92
t_3	516.00	555.92	206.18	0.00	39.92
t_4	510.00	555.92	249.80	0.00	45.92
t_5	515.00	555.92	288.68	0.00	40.92
t_6	544.00	555.92	300.00	0.00	11.92
t_7	646.00	646.00	300.00	0.00	0.00
t_8	686.00	686.00	300.00	0.00	0.00
t_9	741.00	716.28	272.53	24.72	0.00
t_{10}	734.00	716.28	252.83	17.72	0.00
t_{11}	748.00	716.28	217.58	31.72	0.00
t_{12}	760.00	716.28	169.00	43.72	0.00
t_{13}	754.00	716.28	127.08	37.72	0.00
t_{14}	700.00	700.00	127.08	0.00	0.00
t_{15}	686.00	686.00	127.08	0.00	0.00
t_{16}	720.00	716.28	122.95	3.72	0.00
t_{17}	714.00	714.00	122.95	0.00	0.00
t_{18}	761.00	716.28	73.25	44.72	0.00
t_{19}	727.00	716.28	61.34	10.72	0.00
t_{20}	714.00	712.80	60.00	1.20	0.00
t_{21}	618.00	618.00	60.00	0.00	0.00
t_{22}	584.00	584.00	60.00	0.00	0.00
t_{23}	578.00	582.05	63.85	0.00	4.05
t_{24}	544.00	582.05	100.00	0.00	38.05

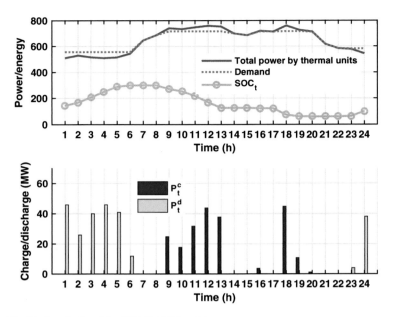

Fig. 7.1 The hourly dispatch of ESS in DED problem

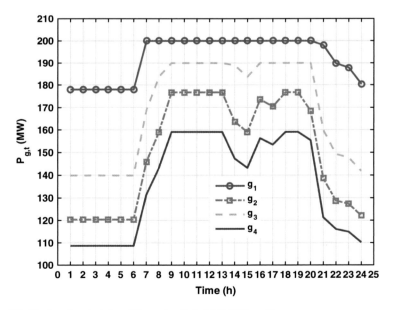

Fig. 7.2 The hourly schedules of thermal units in DED-ESS problem

7.2.2 ESS Operation in Wind-DED Problem

The role of ESS in efficient utilization of wind energy is investigated in this section. The ESS-wind-DED is formulated as follows:

$$\min_{\text{DV}} \text{TC} = \sum_{g,t} a_g P_{g,t}^2 + b_g P_{g,t} + c_g + \sum_t \text{VWC} \times P_t^{\text{wc}} \tag{7.2a}$$

$$\text{DV} = \left\{ P_{g,t}, \text{SOC}_t, P_t^d, P_t^c, P_t^w, P_t^{\text{wc}} \right\} \tag{7.2b}$$

$$P_g^{\min} \le P_{g,t} \le P_g^{\max} \tag{7.2c}$$

$$P_{g,t} - P_{g,t-1} \le \text{RU}_g \tag{7.2d}$$

$$P_{g,t-1} - P_{g,t} \le \text{RD}_g \tag{7.2e}$$

$$\text{SOC}_t = \text{SOC}_{t-1} + \left(P_t^c \eta_c - P_t^d / \eta_d \right) \Delta_t \tag{7.2f}$$

$$P_{\min}^c \le P_t^c \le P_{\max}^c \tag{7.2g}$$

$$P_{\min}^d \le P_t^d \le P_{\max}^d \tag{7.2h}$$

$$\text{SOC}_{\min} \le \text{SOC}_t \le \text{SOC}_{\max} \tag{7.2i}$$

$$P_t^w + \sum_g P_{g,t} + P_t^d \ge L_t - P_t^c \tag{7.2j}$$

$$P_t^w + P_t^{\text{wc}} \le \Lambda_t^w \tag{7.2k}$$

GCode 7.2 The cost-based wind-DED integrated with ESS (7.2)

```
Set      t  /t1*t24/,   g   /p1*p4/;
Table  gendata(g,*)  generator  cost  characteristics  and  limits
         a      b      c    d      e     f    Pmin  Pmax  RU0   RD0
p1    0.12  14.80   89   1.2  −5     3    28   200   40    40
p2    0.17  16.57   83   2.3  −4.24  6.09 20   290   30    30
p3    0.15  15.55  100   1.1  −2.15  5.69 30   190   30    30
p4    0.19  16.21   70   1.1  −3.99  6.2  20   260   50    50;
table  data(t,*)
        lambda  load  wind
t1      32.71   510   44.1
t2      34.72   530   48.5
t3      32.71   516   65.7
t4      32.74   510   144.9
t5      32.96   515   202.3
t6      34.93   544   317.3
t7      44.9    646   364.4
t8      52      686   317.3
t9      53.03   741   271
t10     47.26   734   306.9
t11     44.07   748   424.1
t12     38.63   760   398
```

```
t13    39.91    754    487.6
t14    39.45    700    521.9
t15    41.14    686    541.3
t16    39.23    720    560
t17    52.12    714    486.8
t18    40.85    761    372.6
t19    41.2     727    367.4
t20    41.15    714    314.3
t21    45.76    618    316.6
t22    45.59    584    311.4
t23    45.56    578    405.4
t24    34.72    544    470.4;
* ─────────────────────────────────
variables            cost            Cost of thermal units
                     p(g,t)          Power generated by thermal power
                                     plant
                     SOC(t) ,Pd(t),Pc(t),Pw(t),PWC(t) ;
p.up(g,t) = gendata(g,"Pmax") ; p.lo(g,t) = gendata(g,"Pmin");
scalar SOC0 /100/, SOCmax /300/, eta_c /0.95/, eta_d /0.9/, VWC
      /50/;
SOC.up(t)=SOCmax; SOC.lo(t)=0.2*SOCmax;
SOC.fx('t24')=SOC0; Pc.up(t)=0.2*SOCmax;
Pc.lo(t)=0; Pd.up(t)=0.2*SOCmax; Pd.lo(t)=0;
Pw.up(t)=data(t,'wind'); Pw.lo(t)=0;
Pwc.up(t)=data(t,'wind'); Pwc.lo(t)=0;
Equations Genconst3,Genconst4,costThermalcalc,constESS,balance,
     wind;
costThermalcalc.. cost =e=sum(t,VWC*pwc(t))
  +sum((t,g), gendata(g,'a')*power(p(g,t),2)+gendata(g,'b')*
      p(g,t) +gendata(g,'c'));
Genconst3(g,t) .. p(g,t+1)−p(g,t)=l=gendata(g,'RU0');
Genconst4(g,t) .. p(g,t−1)−p(g,t)=l=gendata(g,'RD0');
constESS(t)..SOC(t)=e=SOC0$(ord(t)=1)+ SOC(t−1)$(ord(t)>1)+Pc(t)*
     eta_c−Pd(t)/eta_d;
balance(t) ..  Pw(t)+ sum(g,p(g,t))+Pd(t)=g=data(t,'load')+Pc(t);
wind(t) ..      Pw(t)+PWC(t)=e=data(t,'wind');
Model DEDESSwind /all/;
Solve DEDESSwind us qcp min cost;
parameter rep(t,*);
rep(t,'Pth')=sum(g,p.l(g,t));
rep(t,'SOC')=SOC.l(t);
rep(t,'Pd')=Pd.l(t);
rep(t,'Pc')=Pc.l(t);
rep(t,'Load')=data(t,'load');
execute_unload "DEDESSwind.gdx" P.l
execute 'gdxxrw.exe DEDESSwind.gdx var=P  rng=Pthermal!a1'
execute_unload "DEDESSwind.gdx" rep
execute 'gdxxrw.exe DEDESSwind.gdx par=reP  rng=rep!a1'
```

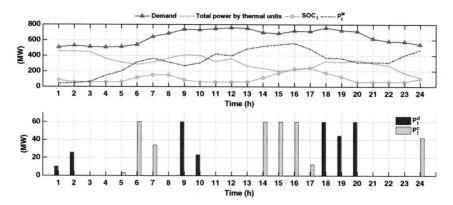

Fig. 7.3 The hourly dispatch of ESS in wind-DED problem

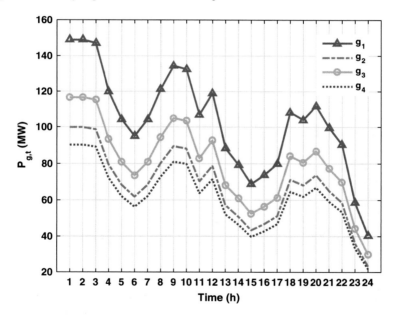

Fig. 7.4 The hourly schedules of thermal units in wind-DED-ESS problem

The total operating costs using ESS are TC = 2.2336×10^5 as obtained in GCode 7.2. The hourly dispatch of ESS in wind-DED problem is depicted in Fig. 7.3. The hourly schedules of thermal units in wind-DED-ESS problem are shown in Fig. 7.4.

7.2.3 ESS Operation in DC-OPF

The multi-period ESS integrated DC-OPF is formulated as follows:

$$\text{OF} = \sum_{g,t} a_g (P_{g,t})^2 + b_g P_{g,t} + c_g + \sum_{i,t} \text{VOLL} \times \text{LS}_{i,t} + \text{VWC} \times P_{i,t}^{\text{wc}} \qquad (7.3a)$$

$$\sum_{g \in \Omega_G^i} P_{g,t} + \text{LS}_{i,t} + P_{i,t}^w - L_{i,t} - P_{i,t}^c + P_{i,t}^d = \sum_{j \in \Omega_\ell^i} P_{ij,t} : \lambda_{i,t} \qquad (7.3b)$$

$$P_{ij,t} = \frac{\delta_{i,t} - \delta_{j,t}}{X_{ij}} \qquad (7.3c)$$

$$-P_{ij}^{\max} \leq P_{ij,t} \leq P_{ij}^{\max} \qquad (7.3d)$$

$$P_g^{\min} \leq P_{g,t} \leq P_g^{\max} \qquad (7.3e)$$

$$P_{g,t} - P_{g,t-1} \leq \text{RU}_g \qquad (7.3f)$$

$$P_{g,t-1} - P_{g,t} \leq \text{RD}_g \qquad (7.3g)$$

$$0 \leq \text{LS}_{i,t} \leq L_{i,t} \qquad (7.3h)$$

$$P_{i,t}^{\text{wc}} = w_{i,t} \Lambda_i^w - P_{i,t}^w \qquad (7.3i)$$

$$0 \leq P_{i,t}^w \leq w_t \Lambda_i^w \qquad (7.3j)$$

$$\text{SOC}_{i,t} = \text{SOC}_{i,t-1} + \left(P_{i,t}^c \eta_c - P_{i,t}^d / \eta_d \right) \Delta_t \qquad (7.3k)$$

$$P_{i,\min}^c \leq P_{i,t}^c \leq P_{i,\max}^c \qquad (7.3l)$$

$$P_{i,\min}^d \leq P_{i,t}^d \leq P_{i,\max}^d \qquad (7.3m)$$

$$\text{SOC}_{i,\min} \leq \text{SOC}_{i,t} \leq \text{SOC}_{i,\max} \qquad (7.3n)$$

The model is implemented on IEEE RTS 24-bus network with ESS integration which is shown in Fig. 7.5.

It is a transmission network with the voltage levels of 138 kV, 230 kV, and Sbase = 100 MVA. The branch data for IEEE RTS 24-bus network is given in Table 7.4 [14]. The from bus, to bus, reactance (X), resistance (r), total line charging susceptance (b), and MVA rating (MVA) are specified in this table. The parallel lines in MATPOWER are merged, and the resultants are given in Table 7.4. The generation data for IEEE RTS 24-bus network is given in Table 7.5. The data of generating units in this network is inspired by Conejo et al. [15] and Bouffard et al. [16] with some modifications. The slack bus is bus 13 in this network.

It is assumed that the network is intact and all branches and generating units are working in normal condition. The SOC of ESS connected to bus 19 and 21 are 200

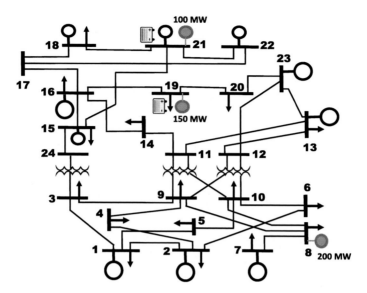

Fig. 7.5 The IEEE 24 RTS with ESS integration

and 100 MW h, respectively. The rest of ESS data is the same as Table 7.1. The minimum operating cost is \$4.1839 × 10^5 obtained by using the GCode 7.3.

GCode 7.3 The ESS-DCOPF GAMS code for IEEE Reliability test 24-bus network

```
Sets bus  /1*24/ ,slack(bus) /13/,Gen /g1*g12/, t /t1*t24/;
scalars Sbase /100/    ,VOLL /10000/,VOLW /50/; alias(bus,node);
Table GD(Gen,*)  Generating units characteristics
      Pmax Pmin   b      CostsD costst RU   RD   SU   SD   UT   DT
            uini U0   So
* The same as before
Set GB(bus,Gen) connectivity index of each generating unit to
    each bus
/* The same as before   / ;
Table BusData(bus,*) Demands of each bus in MW
* The same as before
table branch(bus,node,*)    Network technical characteristics
* The same as before ;
Table WD(t,*)
       w                  d
* The same as before   ;
parameter Wcap(bus)
/8    200
19   150
21   100/;
parameter SOCMax(bus)
/19 200
21   100/;
scalar eta_c /0.95/, eta_d /0.9/, VWC /50/;
parameter SOC0(bus);
```

Table 7.4 Branch data for IEEE RTS 24-bus network

From	To	r(pu)	x(pu)	b(pu)	Rating (MVA)	From	To	r(pu)	x(pu)	b(pu)	Rating (MVA)
1	2	0.0026	0.0139	0.4611	175	11	13	0.0061	0.0476	0.0999	500
1	3	0.0546	0.2112	0.0572	175	11	14	0.0054	0.0418	0.0879	500
1	5	0.0218	0.0845	0.0229	175	12	13	0.0061	0.0476	0.0999	500
2	4	0.0328	0.1267	0.0343	175	12	23	0.0124	0.0966	0.2030	500
2	6	0.0497	0.1920	0.0520	175	13	23	0.0111	0.0865	0.1818	500
3	9	0.0308	0.1190	0.0322	175	14	16	0.0050	0.0389	0.0818	500
3	24	0.0023	0.0839	0.0000	400	15	16	0.0022	0.0173	0.0364	500
4	9	0.0268	0.1037	0.0281	175	15	21	0.0032	0.0245	0.2060	1000
5	10	0.0228	0.0883	0.0239	175	15	24	0.0067	0.0519	0.1091	500
6	10	0.0139	0.0605	2.4590	175	16	17	0.0033	0.0259	0.0545	500
7	8	0.0159	0.0614	0.0166	175	16	19	0.0030	0.0231	0.0485	500
8	9	0.0427	0.1651	0.0447	175	17	18	0.0018	0.0144	0.0303	500
8	10	0.0427	0.1651	0.0447	175	17	22	0.0135	0.1053	0.2212	500
9	11	0.0023	0.0839	0.0000	400	18	21	0.0017	0.0130	0.1090	1000
9	12	0.0023	0.0839	0.0000	400	19	20	0.0026	0.0198	0.1666	1000
10	11	0.0023	0.0839	0.0000	400	20	23	0.0014	0.0108	0.0910	1000
10	12	0.0023	0.0839	0.0000	400	21	22	0.0087	0.0678	0.1424	500

Table 7.5 Generation data for IEEE RTS 24-bus network

Gen	Bus	P_i^{max}	P_i^{min}	b_i ($/MW)	Cs_i ($)	Cd_i ($)	RU_i (MW h^{-1})	RD_i (MW h^{-1})	SU_i (MW h^{-1})	SD_i (MW h^{-1})	UT_i(h)	DT_i(h)	$u_{i,t=0}$	U_i^0 (h)	S_i^0 (h)
g1	18	400	100	5.47	0	0	47	47	105	108	1	1	1	5	0
g2	21	400	100	5.47	0	0	47	47	106	112	1	1	1	6	0
g3	1	152	30.4	13.32	1430.4	1430.4	14	14	43	45	8	4	1	2	0
g4	2	152	30.4	13.32	1430.4	1430.4	14	14	44	57	8	4	1	2	0
g5	15	155	54.25	16	0	0	21	21	65	77	8	8	0	0	2
g6	16	155	54.25	10.52	312	312	21	21	66	73	8	8	1	10	0
g7	23	310	108.5	10.52	624	624	21	21	112	125	8	8	1	10	0
g8	23	350	140	10.89	2298	2298	28	28	154	162	8	8	1	5	0
g9	7	350	75	20.7	1725	1725	49	49	77	80	8	8	0	0	2
g10	13	591	206.85	20.93	3056.7	3056.7	21	21	213	228	12	10	0	0	8
g11	15	60	12	26.11	437	437	7	7	19	31	4	2	0	0	1
g12	22	300	0	0	0	0	35	35	315	326	0	0	1	2	0

```
SOC0( bus )=0.2*SOCMax( bus )/ sbase ;
branch ( bus , node , 'x ' )$( branch ( bus , node , 'x ' )=0)=branch ( node ,
    bus , 'x ' );
branch ( bus , node , 'Limit ' )$( branch ( bus , node , 'Limit ' )=0)=branch
    ( node , bus , 'Limit ' );
branch ( bus , node , 'bij ' )$branch ( bus , node , 'Limit ' ) =1/ branch ( bus ,
    node , 'x ' );
branch ( bus , node , 'z ' )$branch ( bus , node , 'Limit ' )=
sqrt ( power ( branch ( bus , node , 'x ' ) ,2)+power ( branch ( bus , node ,
    'r ' ) ,2 ) );
branch ( node , bus , 'z ' )=branch ( bus , node , 'z ' );
parameter  conex ( bus , node );
conex ( bus , node )$( branch ( bus , node , 'limit ' )and  branch ( node , bus , '
    limit ' ) )=1;
conex ( bus , node )$( conex ( node , bus ) )=1;
Variables  OF, Pij ( bus , node , t ) , Pg ( Gen , t ) , delta ( bus , t ) , lsh ( bus , t ) ,Pw
    ( bus , t ) , pwc ( bus , t ) ,
SOC ( bus , t ) , Pd ( bus , t ) , Pc ( bus , t );
Equations  const1 , const2 , const3 , const4 , const5 , const6 , constESS ;
const1 ( bus , node , t )$(  conex ( bus , node ) )  .. Pij ( bus , node , t )=e=
branch ( bus , node , 'bij ' )*( delta ( bus , t )−delta ( node , t ) );
const2 ( bus , t )  .. lsh ( bus , t )$BusData ( bus , 'pd ' )+Pw ( bus , t )$Wcap ( bus )
+sum ( Gen$GB ( bus , Gen ) , Pg ( Gen , t ) )
+(−Pc ( bus , t )+Pd ( bus , t ) )$SOCMAX ( bus )−WD( t , 'd ' )*BusData ( bus , 'pd ' )/
    Sbase=e=
+sum ( node$conex ( node , bus ) , Pij ( bus , node , t ) );
const3     .. OF=g=sum ( ( bus , Gen , t )$GB ( bus , Gen ) , Pg ( Gen , t )*GD ( Gen , 'b
    ' )*Sbase )
+sum ( ( bus , t ) ,VOLL*lsh ( bus , t )*Sbase$BusData ( bus , 'pd ' )
+VOLW*Pwc ( bus , t )*sbase$Wcap ( bus )  );
const4 ( gen , t )  .. pg ( gen , t+1)−pg ( gen , t )=l=GD ( gen , 'RU ' )/ Sbase ;
const5 ( gen , t )  .. pg ( gen , t−1)−pg ( gen , t )=l=GD ( gen , 'RD ' )/ Sbase ;
const6 ( bus , t )$Wcap ( bus )  .. pwc ( bus , t )=e=WD( t , 'w ' )*Wcap ( bus )/ Sbase
    −pw ( bus , t );
constESS ( bus , t )$SOCMax ( bus ) .. SOC ( bus , t )=e=SOC0 ( bus )$( ord ( t )=1)+
SOC ( bus , t−1)$( ord ( t )>1)+Pc ( bus , t )*eta_c −Pd ( bus , t )/ eta_d ;
Model  loadflow       / all /;
Pg . lo ( Gen , t )=GD ( Gen , 'Pmin ' )/ Sbase ; Pg . up ( Gen , t )=GD ( Gen , 'Pmax ' )/
    Sbase ;
delta . up ( bus , t )=pi /2; delta . lo ( bus , t )=−pi /2; delta . fx ( slack , t )=0;
Pij . up ( bus , node , t )$( ( conex ( bus , node ) ) )= branch ( bus , node , 'Limit ' )/
    Sbase ;
Pij . lo ( bus , node , t )$( ( conex ( bus , node ) ) )=−branch ( bus , node , 'Limit ' )/
    Sbase ;
lsh . up ( bus , t )=WD( t , 'd ' )*BusData ( bus , 'pd ' )/ Sbase ; lsh . lo ( bus , t )=0;
Pw . up ( bus , t )=WD( t , 'w ' )*Wcap ( bus )/ Sbase ; Pw . lo ( bus , t )=0;
Pwc . up ( bus , t )=WD( t , 'w ' )*Wcap ( bus )/ Sbase ; Pwc . lo ( bus , t )=0;
SOC . up ( bus , t )=SOCmax ( bus )/ sbase ; SOC . lo ( bus , t )=0.2*SOCmax ( bus )/
    sbase ;
SOC . fx ( bus , 't24 ' )=SOC0 ( bus );
Pc . up ( bus , t )=0.2*SOCmax ( bus )/ sbase ; Pc . lo ( bus , t )=0;
Pd . up ( bus , t )=0.2*SOCmax ( bus )/ sbase ; Pd . lo ( bus , t )=0;
Solve  loadflow  minimizing OF using lp ;
```

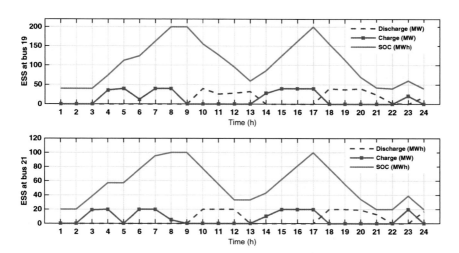

Fig. 7.6 The hourly dispatch of ESS in DC-OPF problem

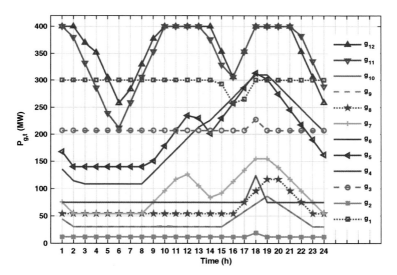

Fig. 7.7 The hourly dispatch of thermal generating units in DC-OPF problem

The hourly dispatch of ESS in DC-OPF problem is shown in Fig. 7.6.

The hourly dispatch of thermal generating units in DC-OPF problem is depicted in Fig. 7.7.

7.2.4 ESS Operation in AC-OPF

The multi-period ESS integrated AC-OPF is formulated as follows:

$$\text{OF} = \sum_{g,t} a_g(P_{g,t})^2 + b_g P_{g,t} + c_g + \sum_{i,t} \text{VOLL} \times \text{LS}_{i,t} + \text{VWC} \times P_{i,t}^{\text{wc}} \tag{7.4a}$$

$$\sum_{g \in \Omega_G^i} P_{g,t} + \text{LS}_{i,t} + P_{i,t}^w - P_{i,t}^l - P_{i,t}^c + P_{i,t}^d = \sum_{j \in \Omega_\ell^i} P_{ij,t} : \lambda_{i,t}^p \tag{7.4b}$$

$$\sum_{g \in \Omega_G^i} Q_{g,t} - Q_{i,t}^l = \sum_{j \in \Omega_\ell^i} P_{ij,t} : \lambda_{i,t}^q \tag{7.4c}$$

$$I_{ij,t} = \frac{V_{i,t}\angle\delta_{i,t} - V_{j,t}\angle\delta_{j,t}}{Z_{ij}\angle\theta_{ij}} + \frac{bV_{i,t}}{2}\angle\left(\delta_{i,t} + \frac{\pi}{2}\right) \tag{7.4d}$$

$$S_{ij,t} = (V_{i,t}\angle\delta_{i,t})I_{ij,t}^* \tag{7.4e}$$

$$P_{ij,t} = \text{real}\{S_{ij,t}\} = \frac{V_{i,t}^2}{Z_{ij}}\cos(\theta_{ij}) - \frac{V_{i,t}V_{j,t}}{Z_{ij}}\cos(\delta_{i,t} - \delta_{j,t} + \theta_{ij}) \tag{7.4f}$$

$$Q_{ij,t} = \text{Img}\{S_{ij,t}\} = \frac{V_{i,t}^2}{Z_{ij}}\sin(\theta_{ij}) - \frac{V_{i,t}V_{j,t}}{Z_{ij}}\sin(\delta_{i,t} - \delta_{j,t} + \theta_{ij}) - \frac{bV_{i,t}^2}{2} \tag{7.4g}$$

$$-S_{ij}^{\max} \leq S_{ij,t} \leq S_{ij}^{\max} \tag{7.4h}$$

$$P_g^{\min} \leq P_{g,t} \leq P_g^{\max} \tag{7.4i}$$

$$Q_g^{\min} \leq Q_{g,t} \leq Q_g^{\max} \tag{7.4j}$$

$$P_{g,t} - P_{g,t-1} \leq \text{RU}_g \tag{7.4k}$$

$$P_{g,t-1} - P_{g,t} \leq \text{RD}_g \tag{7.4l}$$

$$0 \leq \text{LS}_{i,t} \leq p_{i,t}^l \tag{7.4m}$$

$$P_{i,t}^{\text{wc}} = w_{i,t}\Lambda_i^w - P_{i,t}^w \tag{7.4n}$$

$$0 \leq P_{i,t}^w \leq w_t\Lambda_i^w \tag{7.4o}$$

$$\text{SOC}_{i,t} = \text{SOC}_{i,t-1} + \left(P_{i,t}^c\eta_c - P_{i,t}^d/\eta_d\right)\Delta_t \tag{7.4p}$$

$$P_{i,\min}^c \leq P_{i,t}^c \leq P_{i,\max}^c \tag{7.4q}$$

$$P_{i,\min}^d \leq P_{i,t}^d \leq P_{i,\max}^d \tag{7.4r}$$

$$\text{SOC}_{i,\min} \leq \text{SOC}_{i,t} \leq \text{SOC}_{i,\max} \tag{7.4s}$$

The AC-OPF model is implemented on IEEE RTS 24-bus network with ESS integration which is shown in Fig. 7.5. The wind capacities and ESS characteristics are the same as Sect. 7.2.3. The minimum operating cost is 4.3018×10^5 obtained by using the GCode 7.4.

GCode 7.4 The ESS-ACOPF GAMS code for IEEE Reliability test 24-bus network

```
Sets bus  /1*24/ , slack (bus) /13/, Gen /g1*g12/, t /t1*t24/;
Scalars Sbase /100/ , VOLL /10000/, VOLW /50/; alias (bus, node);
Table GD(Gen,*)  Generating units characteristics
set GB(bus, Gen) connectivity index of each generating unit to
    each bus ;
Table BusData(bus,*) Demands of each bus in MW
Table branch (bus, node,*)    Network technical characteristics
              r      x       b      limit;
Table WD(t,*)
Parameter Wcap(bus) Wind capacities at each bus
parameter SOCMax(bus)
/ 19 200
21   100 /; Scalar eta_c /0.95/, eta_d /0.9/, VWC /50/;
Parameter SOC0(bus); SOC0(bus)=0.2*SOCMax(bus)/sbase;
branch(bus,node,'z')$branch(bus,node,'Limit')=sqrt(power(branch
(bus,node,'x'),2)+power(branch(bus,node,'r'),2));
branch(node,bus,'theta')$branch(node,bus,'limit')=
arctan(branch(node,bus,'x')/branch(node,bus,'r')); Parameter
    conex(bus,node);
conex(bus,node)$(branch(bus,node,'limit')and branch(node,bus,'
    limit'))=1;
conex(bus,node)$(conex(node,bus))=1;
Variables OF,Qij(bus,node,t),Pij(bus,node,t),Pg(Gen,t),Qg(Gen,t),
    delta(bus,t)
,V(bus,t),lsh(bus,t),Pw(bus,t),pwc(bus,t),SOC(bus,t),Pd(bus,t),Pc
    (bus,t);
Equations const2,const2B,const3,const4,const5,const6,const7,
    const8,constESS;
const1(bus,node,t)$conex(bus,node).. Pij(bus,node,t)=e=
branch(bus,node,'bij')*(delta(bus,t)−delta(node,t));
const2(bus,t).. lsh(bus,t)$BusData(bus,'pd')+Pw(bus,t)$Wcap(bus)+
    sum(Gen$GB(bus,Gen))
Pg(Gen,t)) +(−Pc(bus,t)+Pd(bus,t))$SOCMAX(bus)−WD(t,'d')*BusData
(bus,'pd')/Sbase=e=+sum(node$conex(node,bus),Pij(bus,node,t));
const2B(bus,t).. sum(Gen$GB(bus,Gen),Qg(Gen,t))−WD(t,'d')*BusData
(bus,'Qd')/Sbase=e=+sum(node$conex(node,bus),Qij(bus,node,t));
const3..OF=g=sum((bus,Gen,t)$GB(bus,Gen),Pg(Gen,t)*GD(Gen,'b')*
    Sbase)+sum((bus,t),
VOLL*lsh(bus,t)*Sbase$BusData(bus,'pd')+VOLW*Pwc(bus,t)*
    sbase$Wcap(bus));
const4(gen,t)  .. pg(gen,t+1)−pg(gen,t)=l=GD(gen,'RU')/Sbase;
const5(gen,t)  .. pg(gen,t−1)−pg(gen,t)=l=GD(gen,'RD')/Sbase;
const6(bus,t)$Wcap(bus) .. pwc(bus,t)=e=WD(t,'w')*Wcap(bus)/Sbase
    −pw(bus,t);
constESS(bus,t)$SOCMax(bus)  .. SOC(bus,t)=e=SOC0(bus)$(ord(t)=1)
                    +SOC(bus,t−1)$(ord(t)>1)+Pc(bus,t)*eta_c−Pd(
                        bus,t)/eta_d;
```

```
const7 (bus , node , t ) $conex ( bus , node ) .. Pij (bus , node , t )=e=
V( bus , t )*V( bus , t )*cos ( branch ( node , bus , ' theta ' )) / branch ( node , bus , '
    z ' )
−V( bus , t )*V( node , t )*cos ( delta ( bus , t )−delta ( node , t )
+branch ( node , bus , ' theta ' )) / branch ( node , bus , ' z ' ) ;
const8 ( bus , node , t ) $conex ( bus , node )  ..  Qij ( bus , node , t )=e=
V( bus , t )*V( bus , t )*sin ( branch ( node , bus , ' theta ' )) / branch ( node , bus , '
    z ' )
−V( bus , t )*V( node , t )*sin ( delta ( bus , t )−delta ( node , t )
+branch ( node , bus , ' theta ' )) / branch ( node , bus , ' z ' )
−  branch ( node , bus , ' b ' )*V( bus , t )*V( bus , t ) / 2 ;  Model  loadflow       /
    all / ;
Pg. lo ( Gen , t )=GD( Gen , ' Pmin ' ) / Sbase ;  Pg. up ( Gen , t )=GD( Gen , ' Pmax ' ) /
    Sbase ;
Qg. lo ( Gen , t )=GD( Gen , ' Qmin ' ) / Sbase ;  Qg. up ( Gen , t )=GD( Gen , ' Qmax ' ) /
    Sbase ;
delta . up ( bus , t )=pi / 2 ;  delta . lo ( bus , t )=−pi / 2 ;  delta . fx ( slack , t )=0 ;
V. up ( bus , t ) = 1.05 ;  V. lo ( bus , t ) = 0.95 ;  V. l ( bus , t ) = 1 ;
Pij . up ( bus , node , t ) $(( conex ( bus , node ) ))=  branch ( bus , node , ' Limit ' ) /
    Sbase ;
Pij . lo ( bus , node , t ) $(( conex ( bus , node ) ))=−branch ( bus , node , ' Limit ' ) /
    Sbase ;
Qij . up ( bus , node , t ) $(( conex ( bus , node ) ))=  branch ( bus , node , ' Limit ' ) /
    Sbase ;
Qij . lo ( bus , node , t ) $(( conex ( bus , node ) ))=−branch ( bus , node , ' Limit ' ) /
    Sbase ;
lsh . up ( bus , t )=  WD( t , ' d ' )*BusData ( bus , ' pd ' ) / Sbase ;  lsh . lo ( bus , t )=
    0 ;
Pw. up ( bus , t )=WD( t , ' w ' )*Wcap ( bus ) / Sbase ;  Pw. lo ( bus , t ) = 0 ;
Pwc. up ( bus , t )=WD( t , ' w ' )*Wcap ( bus ) / Sbase ;  Pwc. lo ( bus , t ) = 0 ;
SOC. up ( bus , t )=SOCmax ( bus ) / sbase ;  SOC. lo ( bus , t ) =0*SOCmax ( bus ) /
    sbase ;
SOC. fx ( bus , ' t24 ' )=SOC0( bus ) ;  Pc. up ( bus , t ) = 0.2 * SOCmax ( bus ) / sbase ;
    Pc. lo ( bus , t ) = 0 ;
Pd. up ( bus , t ) = 0.2 *SOCmax ( bus ) / sbase ;  Pd. lo ( bus , t ) = 0 ;
Solve  loadflow  minimizing  OF  using  nlp ;
```

The hourly dispatch of ESS in AC-OPF problem is shown in Fig. 7.8 and the hourly dispatch of thermal generating units in AC-OPF problem is depicted in Fig. 7.9.

The hourly active power losses in AC-OPF problem are shown in Fig. 7.10. The hourly min/average/max value of voltage magnitudes (pu) is depicted in Fig. 7.11.

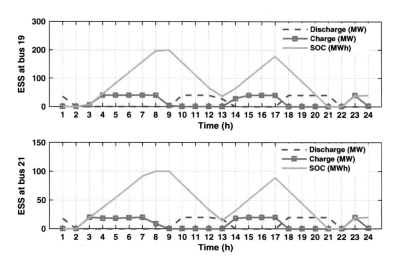

Fig. 7.8 The hourly dispatch of ESS in AC-OPF problem

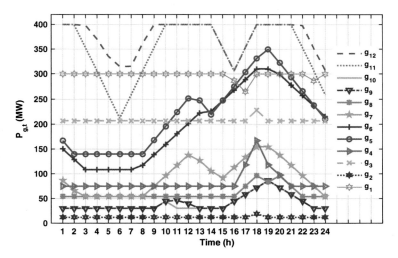

Fig. 7.9 The hourly dispatch of thermal generating units in AC-OPF problem

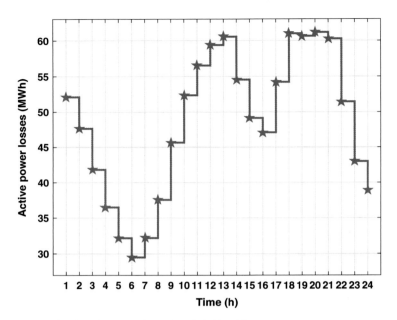

Fig. 7.10 The hourly active power losses in AC-OPF problem

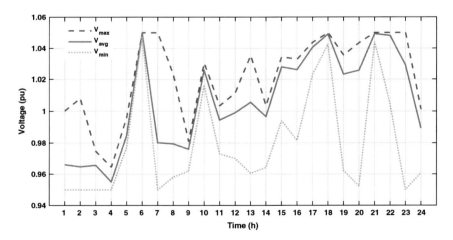

Fig. 7.11 The hourly min/avegare/max values of voltage magnitudes (pu)

7.3   ESS Allocation

The multi-period ESS integrated DC-OPF is formulated as follows:

$$OF = \sum_{g,t} a_g (P_{g,t})^2 + b_g P_{g,t} + c_g + \sum_{i,t} VOLL \times LS_{i,t} + VWC \times P_{i,t}^{wc} \qquad (7.5a)$$

$$\sum_{g\in\Omega_G^i} P_{g,t} + LS_{i,t} + P_{i,t}^w - L_{i,t} - P_{i,t}^c + P_{i,t}^d = \sum_{j\in\Omega_\ell^i} P_{ij,t} : \lambda_{i,t} \qquad (7.5b)$$

$$P_{ij,t} = \frac{\delta_{i,t} - \delta_{j,t}}{X_{ij}} \qquad (7.5c)$$

$$-P_{ij}^{max} \le P_{ij,t} \le P_{ij}^{max} \qquad (7.5d)$$

$$P_g^{min} \le P_{g,t} \le P_g^{max} \qquad (7.5e)$$

$$P_{g,t} - P_{g,t-1} \le RU_g \qquad (7.5f)$$

$$P_{g,t-1} - P_{g,t} \le RD_g \qquad (7.5g)$$

$$0 \le LS_{i,t} \le L_{i,t} \qquad (7.5h)$$

$$P_{i,t}^{wc} = w_{i,t}\Lambda_i^w - P_{i,t}^w \qquad (7.5i)$$

$$0 \le P_{i,t}^w \le w_t\Lambda_i^w \qquad (7.5j)$$

$$SOC_{i,t} = SOC_{i,t-1} + \left(P_{i,t}^c \eta_c - P_{i,t}^d/\eta_d\right)\Delta_t \qquad (7.5k)$$

$$P_{i,min}^c \le P_{i,t}^c \le P_{i,max}^c \qquad (7.5l)$$

$$P_{i,min}^d \le P_{i,t}^d \le P_{i,max}^d \qquad (7.5m)$$

$$SOC_{i,min} \times N_i^{ESS} \le SOC_{i,t} \le SOC_{i,max} \times N_i^{ESS} \qquad (7.5n)$$

$$\sum_i N_i^{ESS} \le N_{max}^{ESS} \qquad (7.5o)$$

Equation (7.5n) enforces the state of charge exist only in those buses that N_i^{ESS} has nonzero value. The total number of ESS units is limited by (7.5o). The model is implemented on IEEE RTS 24-bus network with the wind power generation as shown in Fig. 7.5. The purpose is to find the optimal allocation ESS to minimize the total operating costs. It should be noted that this formulation is only considering 24 h. For planning purposes, longer planning horizon should be taken into account, and load growth and investment costs should also be considered. The ESS allocation is solved using the GCode 7.5 and the total operating cost is obtained as 4.1753×10^5. The optimal locations and capacities of ESS are shown in Fig. 7.12.

GCode 7.5 The ESS allocation using DC-OPF GAMS code for IEEE Reliability test 24-bus network

```
Sets bus  /1*24/ ,slack(bus) /13/,Gen /g1*g12/, t /t1*t24/;
Scalars Nmax , Sbase /100/  ,VOLL /10000/,VWC /50/;  alias(bus,
    node);
Table GD(Gen,*)  Generating units characteristics
* Removed for saving space , the same as previous data
set GB(bus,Gen) connectivity index of each generating unit to
    each bus
/ * Removed for saving space , the same as previous data     / ;
Table BusData(bus,*) Demands of each bus in MW
* Removed for saving space , the same as previous data
table branch(bus,node,*)    Network technical characteristics
                r      x       b       limit
* Removed for saving space , the same as previous data    ;
Table WD(t,*)
* Removed for saving space , the same as previous data    ;
parameter Wcap(bus)
/8     200
19    150
21    100/;
parameter SOCMax(bus); scalar eta_c /0.95/, eta_d /0.9/, VWC
    /50/;
parameter SOC0(bus);
branch(bus,node,'x')$(branch(bus,node,'x')=0)=branch(node,bus,'x'
    );
branch(bus,node,'Limit')$(branch(bus,node,'Limit')=0)=branch(node
    ,bus,'Limit');
branch(bus,node,'bij')$branch(bus,node,'Limit') =1/branch(bus,
    node,'x');
branch(bus,node,'z')$branch(bus,node,'Limit')=sqrt(power(branch
(bus,node,'x'),2)
+power(branch(bus,node,'r'),2));
branch(node,bus,'z')=branch(bus,node,'z');
Parameter conex(bus,node);
conex(bus,node)$(branch(bus,node,'limit')and branch(node,bus,'
    limit'))=1;
conex(bus,node)$(conex(node,bus))=1;
Variables OF, Pij(bus,node,t),Pg(Gen,t),delta(bus,t),lsh(bus,t),Pw
    (bus,t),
pwc(bus,t),SOC(bus,t),Pd(bus,t),Pc(bus,t);
Integer variable NESS(bus);
Equations const1 ,const2 ,const3 ,const4 ,const5 ,const6 ,constESS ,
constESS2 ,constESS3 ,constESS4 ,constESS5 ,constESS6;
const1(bus,node,t)$( conex(bus,node)) .. Pij(bus,node,t)=e=
branch(bus,node,'bij')*(delta(bus,t)−delta(node,t));
const2(bus,t)  .. lsh(bus,t)$BusData(bus,'pd')+Pw(bus,t)$Wcap(bus)
```

```
+sum(Gen$GB(bus,Gen),Pg(Gen,t))+(-Pc(bus,t)+Pd(bus,t))$SOCMAX(bus
        )
-WD(t,'d')*BusData(bus,'pd')/Sbase=e=+sum(node$conex(node,bus),
        Pij(bus,node,t));
const3  ..  OF=g=sum((bus,Gen,t)$GB(bus,Gen),Pg(Gen,t)*GD(Gen,'b')*
        Sbase)+sum((bus,t),
VOLL*lsh(bus,t)*Sbase$BusData(bus,'pd')  +VWC*Pwc(bus,t)*
        sbase$Wcap(bus)));
const4(gen,t)  ..  pg(gen,t+1)-pg(gen,t)=l=GD(gen,'RU')/Sbase;
const5(gen,t)  ..  pg(gen,t-1)-pg(gen,t)=l=GD(gen,'RD')/Sbase;
const6(bus,t)$Wcap(bus)  ..  pwc(bus,t)=e=WD(t,'w')*Wcap(bus)/Sbase
        -pw(bus,t);
constESS(bus,t)    ..  SOC(bus,t)=e=(0.2*NESS(bus)*SOCMax(bus)/sbase
        )$(ord(t)=1)+
SOC(bus,t-1)$(ord(t)>1)+Pc(bus,t)*eta_c-Pd(bus,t)/eta_d;
constESS2(bus,t)  ..  SOC(bus,t)=l=NESS(bus)*SOCmax(bus)/sbase;
constESS3(bus,t)  ..  Pc(bus,t)=l=0.2*NESS(bus)*SOCmax(bus)/sbase;
constESS4(bus,t)  ..  Pd(bus,t)=l=0.2*NESS(bus)*SOCmax(bus)/sbase;
constESS5          ..  sum(bus,NESS(bus))=l=Nmax;
constESS6(bus)  ..  SOC(bus,'t24')=e=0.2*NESS(bus)*SOCmax(bus)/
        sbase;
Model loadflow        /all/;
Pg.lo(Gen,t)=GD(Gen,'Pmin')/Sbase; Pg.up(Gen,t)=GD(Gen,'Pmax')/
        Sbase;
delta.up(bus,t)=pi/2; delta.lo(bus,t)=-pi/2; delta.fx(slack,t)=0;
Pij.up(bus,node,t)$((conex(bus,node)))= branch(bus,node,'Limit')/
        Sbase;
Pij.lo(bus,node,t)$((conex(bus,node)))=-branch(bus,node,'Limit')/
        Sbase;
lsh.up(bus,t)= WD(t,'d')*BusData(bus,'pd')/Sbase; lsh.lo(bus,t)=
        0;
Pw.up(bus,t)=WD(t,'w')*Wcap(bus)/Sbase; Pw.lo(bus,t)=0;
Pwc.up(bus,t)=WD(t,'w')*Wcap(bus)/Sbase; Pwc.lo(bus,t)=0;
SOC.lo(bus,t)=0; Pc.lo(bus,t)=0; Pd.lo(bus,t)=0;
NESS.up(bus)=5; SOCMax(bus)=20; SOC0(bus)=0.2*SOCMax(bus)/sbase;
        Nmax=15;
Solve loadflow minimizing OF using mip;
```

The hourly dispatch of storage unit in optimal ESS allocation problem is shown in Fig. 7.13. The hourly dispatch of thermal generating units in ESS allocation problem is depicted in Fig. 7.14.

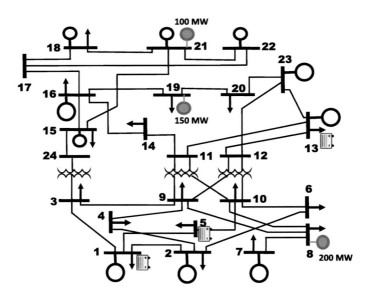

Fig. 7.12 The optimal allocation of ESS units in IEEE RTS network

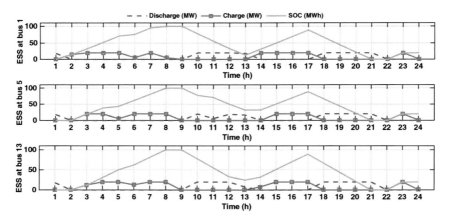

Fig. 7.13 The hourly dispatch of storage unit in optimal ESS allocation problem

Nomenclature

Indices

k Blocks considered for piecewise linear fuel cost function.
i Thermal generating units.
t Time intervals.

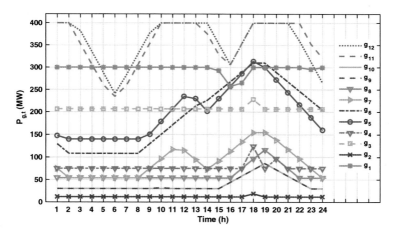

Fig. 7.14 The hourly dispatch of thermal generating units in ESS allocation problem

Parameters

λ_t^e	Electric energy price at time t (\$/MW h).
L_t	Electric demand at time t.
a_g, b_g, c_g	Fuel cost coefficients of unit i.
$P_g^{\max/\min}$	Maximum/minimum limits of power generation of thermal unit i.
RU_g/RD_g	Ramp-up/down limit of generation unit i (MW/h).
VWC	Value of wind curtailment
Λ_t^w	Wind availability at time t

Variables

$U_{i,t}^{\mathrm{ESS}}$	Binary variable indicating the charge/discharge mode of ESS connected to bus i at time t
P_t^w	Wind generation at time t
P_t^{wc}	Wind curtailment at time t
$C_{g,t}$	Fuel cost of thermal unit i at time t (\$).
$P_{g,t}$	Power generated by thermal unit i at time t (MW).
TC	Total operating costs (\$).

7.4 Applications

Various applications of ESS in power system operation and planning studies have been investigated in the literature. Some of these aspects are listed here as follows:

- Electrical Energy Storage and Real-Time Thermal Ratings to Defer Network Reinforcement [17]

- Energy storage and its use with intermittent renewable energy [18]
- Sizing of Energy Storage for microgrids [19]
- Primary Frequency Control using ESS [20]
- OPF in microgrids with ESS [21]
- ESS sizing for grid-connected photovoltaic systems [22]
- Coupling pumped hydro energy storage with UC [23]
- Transmission congestion relief privately owned large-scale ESS [24]
- Phase balancing using ESS in power grids under uncertainty [11]
- Price maker ESS in nodal energy markets [25]
- Value of ESS in dynamic economic dispatch game [26]
- Impact of strategic behavior and ownership of ESS on provision of flexibility [27]
- Reliability improvement using ESS [28]

References

1. A.M. Gee, F. Robinson, W. Yuan, A superconducting magnetic energy storage-emulator/battery supported dynamic voltage restorer. IEEE Trans. Energy Convers. **32**(1), 55–64 (2017)
2. C. Krupke, J. Wang, J. Clarke, X. Luo, Modeling and experimental study of a wind turbine system in hybrid connection with compressed air energy storage. IEEE Trans. Energy Convers. **32**(1), 137–145 (2017)
3. X. Chang, Y. Li, X. Li, X. Chen, An active damping method based on a supercapacitor energy storage system to overcome the destabilizing effect of instantaneous constant power loads in dc microgrids. IEEE Trans. Energy Convers. **32**(1), 36–47 (2017)
4. B. Steffen, Prospects for pumped-hydro storage in germany. Energy Policy **45**, 420–429 (2012)
5. J.M. Lujano-Rojas, R. Dufo-Lpez, J.L. Bernal-Agustn, J.P.S. Catalo, Optimizing daily operation of battery energy storage systems under real-time pricing schemes. IEEE Trans. Smart Grid **8**(1), 316–330 (2017)
6. M.I. Daoud, A.M. Massoud, A.S. Abdel-Khalik, A. Elserougi, S. Ahmed, A flywheel energy storage system for fault ride through support of grid-connected VSC HVDC-based offshore wind farms. IEEE Trans. Power Syst. **31**(3), 1671–1680 (2016)
7. C. Park, R. Sedundo, V. Knazkins, P. Korba, Feasibility analysis of the power-to-gas concept in the future swiss power system, in: *CIRED Workshop 2016* (2016), pp. 1–5
8. S. Pulendran, J.E. Tate, Energy storage system control for prevention of transient under-frequency load shedding. IEEE Trans. Smart Grid **8**(2), 927–936 (2017)
9. C.L.T. Borges, D.M. Falcao, Optimal distributed generation allocation for reliability, losses and voltage improvement. Int. J. Electr. Power Energy Syst. **28**(6), 413–420 (2006)
10. P. Maghouli, A. Soroudi, A. Keane, Robust computational framework for mid-term techno-economical assessment of energy storage. IET Gener. Transm. Distrib. **10**(3), 822–831 (2016)
11. S. Sun, B. Liang, M. Dong, J.A. Taylor, Phase balancing using energy storage in power grids under uncertainty. IEEE Trans. Power Syst. **31**(5), 3891–3903 (2016)
12. L.S. Vargas, G. Bustos-Turu, F. Larran, Wind power curtailment and energy storage in transmission congestion management considering power plants ramp rates. IEEE Trans. Power Syst. **30**(5), 2498–2506 (2015)
13. A. Soroudi, P. Siano, A. Keane, Optimal DR and ESS scheduling for distribution losses payments minimization under electricity price uncertainty. IEEE Trans. Smart Grid **7**(1), 261–272 (2016)

14. R.D. Zimmerman, C.E. Murillo-Sanchez, R.J. Thomas, Matpower: steady-state operations, planning, and analysis tools for power systems research and education. IEEE Trans. Power Syst. **26**(1), 12–19 (2011)
15. A.J. Conejo, M. Carrión, J.M. Morales, *Decision Making Under Uncertainty in Electricity Markets*, vol. 1 (Springer, New York, 2010)
16. F. Bouffard, F.D. Galiana, A.J. Conejo, Market-clearing with stochastic security-part II: case studies. IEEE Trans. Power Syst. **20**(4), 1827–1835 (2005)
17. D.M. Greenwood, N.S. Wade, P.C. Taylor, P. Papadopoulos, N. Heyward, A probabilistic method combining electrical energy storage and real-time thermal ratings to defer network reinforcement. IEEE Trans. Sustain. Energy **8**(1), 374–384 (2017)
18. J.P. Barton, D.G. Infield, Energy storage and its use with intermittent renewable energy. IEEE Trans. Energy Convers. **19**(2), 441–448 (2004)
19. S.X. Chen, H.B. Gooi, M.Q. Wang, Sizing of energy storage for microgrids. IEEE Trans. Smart Grid **3**(1), 142–151 (2012)
20. A. Oudalov, D. Chartouni, C. Ohler, Optimizing a battery energy storage system for primary frequency control. IEEE Trans. Power Syst. **22**(3), 1259–1266 (2007)
21. Y. Levron, J.M. Guerrero, Y. Beck, Optimal power flow in microgrids with energy storage. IEEE Trans. Power Syst. **28**(3), 3226–3234 (2013)
22. Y. Ru, J. Kleissl, S. Martinez, Storage size determination for grid-connected photovoltaic systems. IEEE Trans. Sustain. Energy **4**(1), 68–81 (2013)
23. K. Bruninx, Y. Dvorkin, E. Delarue, H. Pandi, W. Dhaeseleer, D.S. Kirschen, Coupling pumped hydro energy storage with unit commitment. IEEE Trans. Sustain. Energy **7**(2), 786–796 (2016)
24. H. Khani, M.R. Dadash Zadeh, A.H. Hajimiragha, Transmission congestion relief using privately owned large-scale energy storage systems in a competitive electricity market. IEEE Trans. Power Syst. **31**(2), 1449–1458 (2016)
25. H. Mohsenian-Rad, Coordinated price-maker operation of large energy storage units in nodal energy markets. IEEE Trans. Power Syst. **31**(1), 786–797 (2016)
26. W. Tang, R. Jain, Dynamic economic dispatch game: the value of storage. IEEE Trans. Smart Grid **7**(5), 2350–2358 (2016)
27. K. Hartwig, I. Kockar, Impact of strategic behavior and ownership of energy storage on provision of flexibility. IEEE Trans. Sustain. Energy **7**(2), 744–754 (2016)
28. A. Vieira Pombo, J. Murta-Pina, V. Ferno Pires, Multiobjective formulation of the integration of storage systems within distribution networks for improving reliability. Electr. Power Syst. Res. **148**, 87–96 (2017)

Chapter 8
Power System Observability

This chapter provides a solution for increasing the power system observability by allocation of Phasor Measurement Units (PMU) problem in GAMS. The PMU is able to measure the voltage phasor at the connection bus and also it measures the current phasor of any branch connected to the bus hosting the PMU. Different cases are analyzed and the problem is tested on some standard IEEE cases.

8.1 Min No. PMU Placement

The PMU devices are able to provide the system operator a set of synchronized phasor measurements in the system in order to make it observable. In this case, it is needed to make every existing bus observable no matter if it has any generation/demand or not. Additionally, the contingencies are not considered. In order to explain the observability concept, Fig. 8.1 provides an example. If a PMU is installed on bus i then bus i and all adjacent buses are observable. It means that the voltage phasors of these buses are known. The system is called observable if every bus of the network is observable.

The PMU placement concept is simply explained as follows: If a bus or at least one of its adjacent buses is equipped with PMU then this bus is observable. One trivial solution for PMU placement problem is installing PMU at all buses. This solution makes all buses observable but obviously the costs would be very high and sometimes practical aspects do not allow this to happen. In other words, it is tried to find the minimum number of buses to have PMU devices in order to make the whole system observable. This PMU placement problem is formulated as follows:

$$\min_{\chi_i} \mathrm{OF} = \sum_{i \in \Omega_B} \chi_i \tag{8.1a}$$

© Springer International Publishing AG 2017
A. Soroudi, *Power System Optimization Modeling in GAMS*,
DOI 10.1007/978-3-319-62350-4_8

Fig. 8.1 PMU functionality
and the observability concept

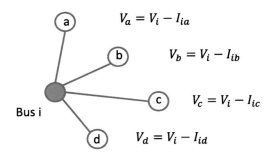

$$\chi_i + \sum_{j\in\Omega_{ij}^\ell} \chi_j \ge \alpha_i \ \forall i \in \Omega_B \tag{8.1b}$$

$$\alpha_i = 1 \tag{8.1c}$$

where i, j are bus index, χ_i is a binary variable indicating if it has PMU (1) or not (0), Ω_B is the set of all network buses, α_i is a binary parameter (observability) which is set to 1 to make the whole network observable, and Ω_i is the set of all buses adjacent (connected) to bus i.

The GAMS code for solving the optimal PMU allocation in IEEE 14-bus network without considering the zero bus injection is provided in GCode 8.1.

GCode 8.1 PMU allocation for IEEE 14 network without considering zero injection nodes

```
Sets  bus    /1*14/;

Alias ( bus , node ) ;

Table  BusData ( bus , * )  buss  characteristics
            Pd        gen
1           0         1
2           21.7      1
3           94.2      1
4           47.8
5           7.6
6           11.2      1
8           0         1
9           29.5
10          9
11          3.5
12          6.1
13          13.5
14          14.9      ;
set  conex                Bus  connectivity  matrix
```

```
/ 1          .          2
1            .          5
2            .          3
2            .          4
2            .          5
3            .          4
4            .          5
4            .          7
4            .          9
5            .          6
6            .          11
6            .          12
6            .          13
7            .          8
7            .          9
9            .          10
9            .          14
10           .          11
12           .          13
13           .          14/;
conex(bus, node)$(conex(node, bus))=1;
Variables  OF;
Binary  variable  PMU(bus);
Equations
const1,
const2;

const1      ..  OF=g=sum(bus ,PMU(bus));
const2(bus)  ..  PMU(bus)+sum(node$conex(bus, node) ,PMU(
    node))=g=1;

Model  placement          /const1 ,const2/;

Solve  placement  minimizing  OF using  mip;
```

The optimal PMU allocation for IEEE 14-bus without considering the zero injection buses is obtained using GCode 8.1 and is shown in Fig. 8.2. The simulation results show that four PMU devices are required to make the IEEE 14-bus network fully observable. The optimal PMU allocation for IEEE 57-bus without considering the zero injection buses is shown in Fig. 8.3. The simulation results show that 14 PMU devices are needed to make the IEEE 57-bus network fully observable.

The optimal PMU allocation for IEEE 118-bus without considering the zero injection buses is shown in Fig. 8.4. The simulation results show that 32 PMU devices are required to make the IEEE 118-bus network fully observable.

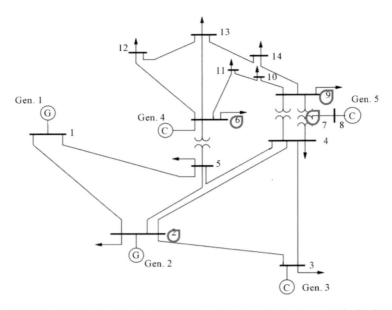

Fig. 8.2 Optimal PMU allocation for IEEE 14-bus without considering the zero injection buses

PMU location

1
6
9
15
19
22
25
27
29
32
36
39
41
45
47
50
53

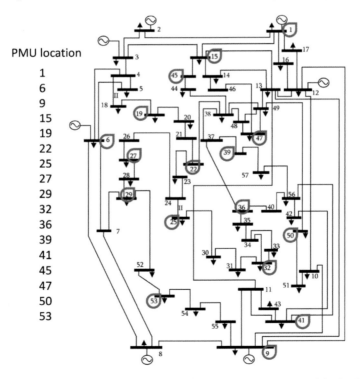

Fig. 8.3 Optimal PMU allocation for IEEE 57-bus without considering the zero injection buses

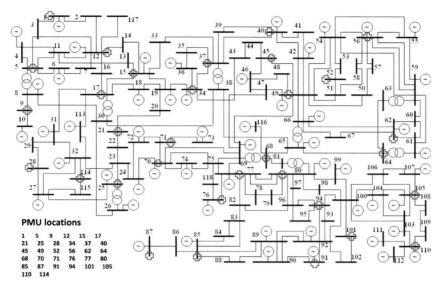

Fig. 8.4 Optimal PMU allocation for IEEE 118-bus without considering the zero injection buses

8.2   Min Cost PMU placement

The cost of PMU device is basically dependent on the number of channels used for measurement. In [1], it is stated that if the cost of each PMU is C and the number of used channels is n at bus i then the total PMU cost would be $w_i = (1 + 0.1 \times n)C$ in the mentioned bus. n is the number of branches connected to bus i.

$$\min_{\chi_i} \text{OF} = \sum_{i \in \Omega_B} w_i \chi_i \tag{8.2a}$$

$$\chi_i + \sum_{j \in \Omega_{ij}^\ell} \chi_j \geq \alpha_i \ \ \forall i \in \Omega_B \tag{8.2b}$$

$$\alpha_i = 1 \tag{8.2c}$$

where i, j are bus index, χ_i is a binary variable indicating if it has PMU (1) or not (0), Ω_B is the set of all network buses, and α_i is a binary parameter (observability) which is set to 1 to make the whole network observable.

The GAMS code for solving the optimal allocation of PMU devices with minimum cost without considering the ZIB is described in GCode 8.2.

GCode 8.2 Min cost PMU allocation for IEEE 14 network without considering zero injection nodes

```
Sets  bus    /1*14/;
alias(bus,node);
Set  conex              Bus  connectivity  matrix
/1          .      2
1           .      5
2           .      3
2           .      4
2           .      5
3           .      4
4           .      5
4           .      7
4           .      9
5           .      6
6           .      11
6           .      12
6           .      13
7           .      8
7           .      9
9           .      10
9           .      14
10          .      11
12          .      13
13          .       14/;
conex(bus,node)$(conex(node,bus))=1;
parameters  cost(bus);
cost(bus)=1+0.1*sum(node$conex(bus,node),1);
Variables
OF_c;
binary  variable
PMU(bus);
Equations
const0 ,
const2;
const0        ..  OF_c=e=sum(bus,cost(bus)*PMU(bus));
const2(bus)  ..  PMU(bus)+sum(node$conex(bus,node),
    PMU(node))=g=1;

Option  optcr=0;

Model  placement0        /const0 ,const2/;
Solve  placement0  minimizing  OF_c  using  mip;
Display  pmu.1;
```

Network	PMU locations
IEEE 14	2 , 8 , 10 , 13
IEEE 57	2 , 6 , 12 , 19 , 22 , 25
	27 , 32 , 36 , 39 , 41 , 45
	46 , 47 , 50 , 52 , 55
IEEE 118	1 , 5 , 10 , 12 , 15 , 17
	21 , 23 , 28 , 30 , 36 , 40
	44 , 46 , 50 , 52 , 56 , 62
	64 , 71 , 75 , 77 , 80 , 85
	87 , 91 , 94 , 101 , 105 , 110
	114 , 116

Table 8.1 Optimal allocation of PMU devices with minimum cost without considering the ZIB

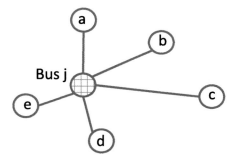

Fig. 8.5 PMU functionality and observability concept in zero injection buses

The minimum cost PMU allocation problem is solved and the solutions are given in Table 8.1.

8.3 Min No. PMU Placement Considering ZIB

Some buses in the network do not contain demand or generation which are called Zero Injection Buses (ZIB). These buses can be specially treated in terms of observability as explained in Fig. 8.5. This figure shows zero injection bus j and its adjacent buses. The observability of each bus is guaranteed with the observability of $n - 1$ of them [2]. In this way, the network buses Ω_B will be divided into two categories as follows:

- Zero injection buses Ω_z or adjacent to zero injection buses Ω_{za}
- The rest of the buses (normal buses) Ω_n

$$\min_{\chi_i} \text{OF} = \sum_{i \in \Omega_B} \chi_i \tag{8.3a}$$

Table 8.2 ZIB buses in different networks under study

Network	Zero injection buses
IEEE 14	7
IEEE 57	4,7,11,21,22,24,26,34
	36,37,39,40,45,46,48
IEEE 118	5,9,30,37,38,63,64,68,71,81

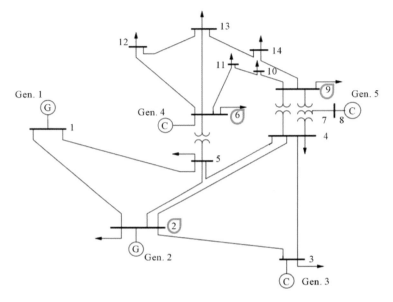

Fig. 8.6 Optimal PMU allocation for IEEE 14-bus with considering the zero injection buses

$$\chi_i + \sum_{j \in \Omega_{ij}^\ell} \chi_j \geq \alpha_i \ \forall i \in \Omega_n \tag{8.3b}$$

$$\sum_{a \in \Omega_{ia}^\ell} \left\langle \chi_a + \sum_{j \in \Omega_{aj}^\ell} \chi_j \right\rangle \geq |\Omega_{za}| - 1 \ \forall i \in \Omega_{za} \tag{8.3c}$$

$$\alpha_i = 1 \tag{8.3d}$$

where i, j are bus index, χ_i is a binary variable indicating if it has PMU (1) or not (0), Ω_B is the set of all network buses, and α_i is a binary parameter (observability) which is set to 1 to make the whole network observable.

The zero injection buses for the networks under study are specified in Table 8.2. The GAMS code for solving the optimal PMU allocation in IEEE 14-bus network without considering the zero bus injection is provided in GCode 8.1 The optimal PMU allocation for IEEE 14 and 57 bus systems with considering the zero injection buses are given in Figs. 8.6 and 8.7, respectively. The optimal PMU allocation for IEEE 118-bus with considering the zero injection buses is shown in Fig. 8.8. The

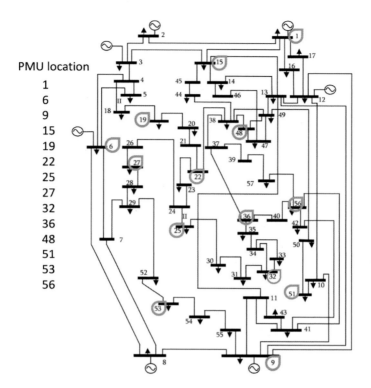

PMU location

1
6
9
15
19
22
25
27
32
36
48
51
53
56

Fig. 8.7 Optimal PMU allocation for IEEE 57-bus with considering the zero injection buses

simulation results show that 29 PMU devices are required to make the IEEE 118-bus network fully observable.

GCode 8.3 PMU allocation for IEEE 14 network considering zero injection nodes

```
Set  bus    /1*14/;  alias(bus,node);  alias(bus,shin);
Table  BusData(bus,*)  bus  characteristics
             Pd           gen
1            0            1
2            21.7         1
3            94.2         1
4            47.8
5            7.6
6            11.2         1
8            0            1
9            29.5
10           9
11           3.5
12           6.1
13           13.5
```

```
14          14.9        ;
set conex            Bus connectivity matrix
/1       .      2
1        .      5
2        .      3
2        .      4
2        .      5
3        .      4
4        .      5
4        .      7
4        .      9
5        .      6
6        .      11
6        .      12
6        .      13
7        .      8
7        .      9
9        .      10
9        .      14
10       .      11
12       .      13
13       .      14/;
conex(bus,node)$(conex(node,bus))=1;
parameters A(bus,node),B(bus,node),Z(bus),normal(bus),
    abnormal(bus);
Z(bus)$(BusData(bus,'Pd')=0 and BusData(bus,'gen')=0)=
    yes;
A(bus,node)$(Z(bus) and conex(bus,node))=yes;
normal(bus)$( Z(bus)=0 and sum(node$conex(bus,node),
Z(node))=0)=yes;
B(bus,node)$(A(bus,node))=yes;
B(bus,node)$(Z(node) and ord(bus)=ord(node))=yes;
abnormal(bus)$(normal(bus)=0)=yes; Variables OF; binary
    variable PMU(bus);
Equations const1,eq1,eq2;
const1     .. OF=g=sum(bus,PMU(bus));
eq1(bus)$normal(bus) .. PMU(bus)+sum(node$conex(bus,
    node),PMU(node))=g=1;
eq2(bus)$abnormal(bus) ..    sum(shin$conex(bus,shin),
PMU(shin)+sum(node$conex(shin,node),PMU(node)) ) =g=sum
    (node,B(bus,node))-1;
Model placementWithzeroinjection    /const1,eq1,eq2/;
Solve placementWithzeroinjection minimizing OF using
    mip;
```

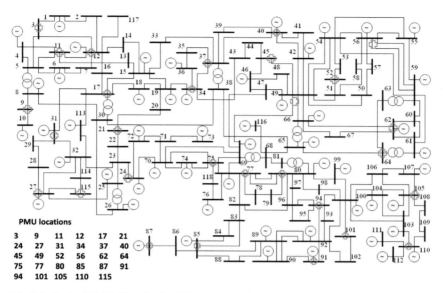

PMU locations

3	9	11	12	17	21
24	27	31	34	37	40
45	49	52	56	62	64
75	77	80	85	87	91
94	101	105	110	115	

Fig. 8.8 Optimal PMU allocation for IEEE 118-bus with considering the zero injection buses

8.4 Min No. PMU Placement for Maximizing the Observability

Another question that can be asked is how to maximize the system observability using a limited number of PMU devices. How should these units be allocated in a given system? In order to answer this question, a binary variable α_i is defined which states bus i is observable (1) or not (0). The optimization problem to be solved is as follows:

$$\max_{\chi_i,\alpha_i} \text{OF} = \sum_{i \in \Omega_B} \alpha_i \tag{8.4a}$$

$$\sum_{i \in \Omega_B} \chi_i \leq N_{\text{PMU}} \tag{8.4b}$$

$$\chi_i + \sum_{j \in \Omega_{ij}^\ell} \chi_j \geq \alpha_i \ \forall i \in \Omega_B \tag{8.4c}$$

$$1 \leq \alpha_i \tag{8.4d}$$

where i, j are bus index, χ_i is a binary variable indicating if it has PMU (1) or not (0), Ω_B is the set of all network buses, α_i is an integer variable (observability) which its minimum value is set to 1 to make the whole network observable, and N_{PMU} is the number of available PMU devices.

Table 8.3 Optimal allocation of PMU devices for maximizing the system observability of IEEE 14-bus network

Bus	N_{PMU}			
	1	2	3	4
2			1	1
4	1			
6		1	1	
7				1
9		1	1	
10				1
13				1
Total observable buses	6	10	13	14

The GAMS code for solving the optimal PMU allocation in IEEE 14-bus network for maximizing the system observability is provided in GCode 8.4. The optimal solution is described in Table 8.3.

GCode 8.4 PMU allocation for IEEE 14 network for maximizing the system observability

```
Sets  bus   /1*14/;  alias(bus,node);  alias(bus,shin);
Table  BusData(bus,*)  buss  characteristics
         Pd        gen
1        0         1
2        21.7      1
3        94.2      1
4        47.8
5        7.6
6        11.2      1
8        0         1
9        29.5
10       9
11       3.5
12       6.1
13       13.5
14       14.9      ;
set  conex            Bus  connectivity  matrix
/1        .       2
1         .       5
2         .       3
2         .       4
2         .       5
3         .       4
4         .       5
4         .       7
4         .       9
5         .       6
```

```
6              .        11
6              .        12
6              .        13
7              .        8
7              .        9
9              .        10
9              .        14
10             .        11
12             .        13
13             .        14/;
conex ( bus , node )$( conex ( node , bus ) ) =1;
parameters A( bus , node ) ,B( bus , node ) ,Z( bus ) , normal ( bus ) ,
    abnormal ( bus ) ;
Z( bus )$( BusData ( bus , 'Pd' ) =0  and  BusData ( bus , 'gen' ) =0)=
    yes ;
A( bus , node )$(Z( bus )  and  conex ( bus , node ) ) =yes ;
normal ( bus )$(  Z( bus ) =0  and  sum( node$conex ( bus , node ) ,
Z( node ) ) =0)=yes ;
abnormal ( bus )$( normal ( bus ) =0)=yes ;  B( bus , node )$(A( bus ,
    node ) ) =yes ;
B( bus , node )$(Z( node )  and  ord ( bus ) =ord ( node ) ) =yes ;
Variables OF, OBI ; Binary  variables PMU( bus ) , alpha ( bus )
    ; Scalar NPMU / 1 0/;
Equations eq3 , eq4 , eq5   ;
eq3 ..    sum( bus , pmu ( bus ) ) =1=NPMU;  eq4 ..  OBI=e=sum
( node , alpha ( node ) ) ;
eq5 ( bus )  ..  PMU( bus )+sum( node$conex ( bus , node ) ,PMU( node )
    )=g= alpha ( bus ) ;
Option optcr =0; Model placement3       / eq3 , eq4 , eq5 /;
Solve placement3 maximizing OBI using mip;
set counter / c1*c4 /; parameter report ( bus , counter ) ,
    OBIrep ( counter ) ;
loop ( counter , NPMU=ord ( counter ) ; Solve placement3
    maximizing OBI using mip;
report ( bus , counter )=pmu . l ( bus ) ; OBIrep ( counter )=OBI. l ;
    ) ;
```

The total observable buses vs N_{PMU} in IEEE 57-bus network is shown in Fig. 8.9.
The total observable buses vs N_{PMU} in IEEE 118-bus network is shown in Fig. 8.10.

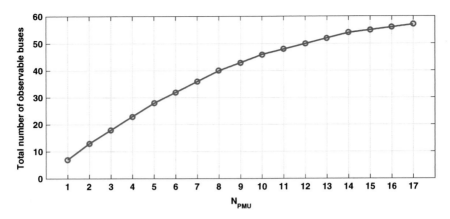

Fig. 8.9 Total observable buses vs N_{PMU} in IEEE 57-bus network

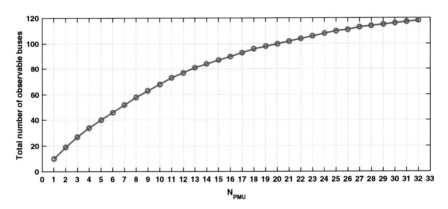

Fig. 8.10 Total observable buses vs N_{PMU} in IEEE 118-bus network

8.5   Min No. PMU Placement Considering Contingencies

The system observability should be maintained in normal operating condition as well as in the case of contingencies. The analyzed contingencies include:

- Loss of single PMU
- Loss of single line

8.5.1   Loss of PMU

In order to make the system observability resilient against the loss of single PMU device, each bus should be observed at least by two PMU devices. The PMU allocation resilient against single PMU outage is formulated as follows:

$$\min_{\chi_i} OF = \sum_{i \in \Omega_B} \chi_i \tag{8.5a}$$

$$\chi_i + \sum_{j \in \Omega_{ij}^{\ell}} \chi_j \geq 2\alpha_i \quad \forall i \in \Omega_B \tag{8.5b}$$

$$\alpha_i = 1 \tag{8.5c}$$

α_i is a binary parameter which shows the observability of each bus and is set to 1 to make the whole network observable. The GAMS code for PMU allocation of IEEE 14 network resilient against single PMU outage is provided in GCode 8.5.

GCode 8.5 PMU allocation for IEEE 14 network resilient against single PMU outage

```
Sets
bus    /1*14/;
alias(bus,node);
alias(bus,shin);
set  conex            Bus connectivity matrix
/1          .           2
1           .           5
2           .           3
2           .           4
2           .           5
3           .           4
4           .           5
4           .           7
4           .           9
5           .           6
6           .           11
6           .           12
6           .           13
7           .           8
7           .           9
9           .           10
9           .           14
10          .           11
12          .           13
13          .           14/;
```

```
conex ( bus , node ) $ ( conex ( node , bus ) ) = 1 ;
Variables
OF , OBI ;

binary  variable
PMU( bus ) ;

Equations
const1 ,
const2A  ;
const1       .. OF=e=sum( bus ,PMU( bus ) ) ;
const2A ( bus )  .. PMU( bus )+sum ( node$conex ( bus , node ) ,PMU
( node ) )=g=2 ;

Option  optcr =0 ;

Model  placement4        / const1 , const2A / ;

Solve  placement4  minimizing  OF  using  mip ;

Display  pmu . l ;
```

The optimal PMU allocation for IEEE 14-bus resilient against single PMU outage
is shown in Fig. 8.11. As it can be seen in this figure, 9 PMU devices are needed
to make the system fully observable even in the case of single PMU outage. The
optimal PMU allocation for IEEE 57-bus resilient against single PMU outage is
shown in Fig. 8.12. As it can be seen in this figure, 33 PMU devices are needed to
make the system fully observable even in the case of single PMU outage.

The optimal PMU allocation for IEEE 118-bus resilient against single PMU
outage is shown in Fig. 8.13. As it can be seen in this figure, 68 PMU devices are
needed to make the system fully observable even in the case of single PMU outage.

8.5.2 Single Line Outage

In this section, PMU units are allocated in the network in order to keep the system
observable even in case of single line outage. For this purpose, each bus should
have a PMU or should be monitored with at least two other PMU devices. This is
formulated as follows:

$$\min_{\chi_i, \alpha_i} OF = \sum_{i \in \Omega_B} \chi_i \tag{8.6a}$$

$$\chi_i + \alpha_i \geq 1 \quad \forall i \in \Omega_B \tag{8.6b}$$

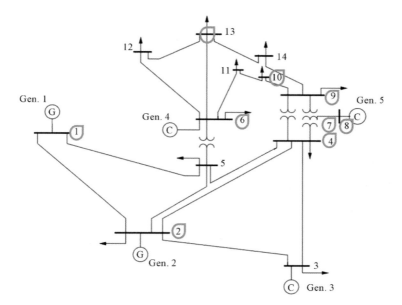

Fig. 8.11 Optimal PMU allocation for IEEE 14-bus resilient against single PMU outage

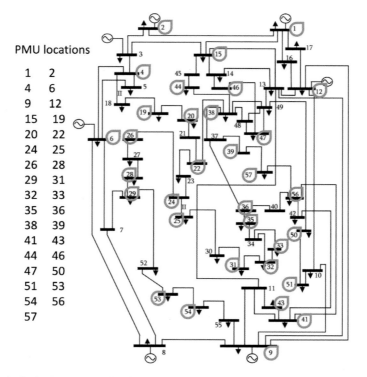

Fig. 8.12 Optimal PMU allocation for IEEE 57-bus resilient against single PMU outage

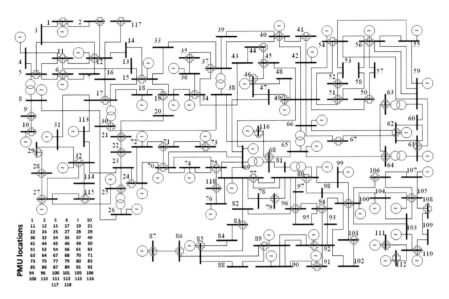

Fig. 8.13 Optimal PMU allocation for IEEE 118-bus resilient against single PMU outage

$$\sum_{j\in\Omega_{ij}^{\ell}} \chi_j \geq 2 \times \alpha_i \quad \forall i \in \Omega_B \tag{8.6c}$$

$$\alpha_i, \chi_i \in \{0, 1\} \tag{8.6d}$$

The GAMS code for solving the PMU allocation considering single line outage is developed in GCode 8.6. α_i is a binary variable which indicates if bus i is monitored using the PMU installed in the adjacent buses. The optimal solution is given in Table 8.4.

GCode 8.6 PMU allocation for IEEE 14 network resilient against single line outage

Sets bus	/1*14/;	Alias(bus,node);
set conex		Bus connectivity matrix
/1	.	2
1	.	5
2	.	3
2	.	4
2	.	5
3	.	4
4	.	5
4	.	7
4	.	9
5	.	6

```
6            .          11
6            .          12
6            .          13
7            .          8
7            .          9
9            .          10
9            .          14
10           .          11
12           .          13
13           .          14/;
conex(bus,node)$(conex(node,bus))=1;
Variables OF; Binary variable PMU(bus),alpha(bus)  ;
scalar NPMU /10/; Equations const1, const2C, const2D;
const1      .. OF=e=sum(bus,PMU(bus));
const2C(bus)  .. PMU(bus)+alpha(bus)=g=1;
const2D(bus)  .. sum(node$conex(bus,node),PMU(node))=g
   =2*alpha(bus);
Option optcr=0;
Model placement6        /const1,const2C,const2D/;
Solve placement6 minimizing OF using mip;
Display pmu.l;
```

Table 8.4 Optimal allocation of PMU devices considering single line outage

Network	Optimal locations
IEEE 14	1,3,6,8,9,10,13
IEEE 57	1,3,5,8,11,12,15,18,20,22,24,27
	29,30,32,33,35,36,39,41,44,46,48
	49,51,53,55,56
IEEE 118	12,15,17,19,21,23,26,27,29,32
	34,36,37,38,40,42,44,46,49,50
	51,53,56,59,61,64,67,68,72,73
	74,75,76,78,80,83,85,87,89,91
	92,94,96,100,102,105,106,109
	111,112,114,116,117

8.6 Multistage PMU Placement

The optimal number of PMU units for making the system fully observable is obtained so far in previous sections. The next step is the implementation of these plans by the system operator. The ideal situation is installing all PMU units in the obtained network buses. However, simultaneous installation of theses devices is not possible due to some practical difficulties such as unavailability of crews and difficulties in simultaneous maintenance outage of lines for PMU installations. In this section, it is assumed that the optimal number and locations of PMU units are already obtained. The system operator can only install a limited number of PMU devices at a given period of time. The question is that which buses should be equipped with PMU at each phase. This problem is called Multistage PMU placement. The decision maker should maximize the number of observable buses of the systems at each phase. The number of observable buses will increase in each phase and after the final phase, the whole system will be observable. In order to solve the problem, the following steps should be taken:

1. The optimal number of PMU units are found using the techniques discussed so far. It might be considering ZIP or contingencies or costs. Save this number N_{pmu} and the locations of PMU devices.
2. How many installation stages and how many PMU units (ψ_p) are going to be installed in each stage?
3. The following optimization problem is solved for stage p.

$$\max_{\chi_i, \alpha_i} \text{OBI} = \sum_{i \in \Omega_B} \alpha_i \tag{8.7a}$$

$$\text{TPMU} = \sum_{i \in \Omega_B} \chi_i \leq \psi_p \tag{8.7b}$$

$$\chi_i + \sum_{j \in \Omega_{ij}^\ell} \chi_j \geq \alpha_i \ \forall i \in \Omega_B \tag{8.7c}$$

$$\chi_i, \alpha_i \in \{0, 1\} \tag{8.7d}$$

It is trying to maximize the number of observable buses in the network subject to number of available installable PMU devices.

The GAMS code for solving the multistage optimal PMU allocation for IEEE 14-bus network is provided in GCode 8.7.

GCode 8.7 Multi-stage PMU allocation for IEEE 14 network

```
Sets
bus    /1*14/;
alias (bus , node );
set  conex              Bus  connectivity  matrix
/1        .        2
1         .        5
2         .        3
2         .        4
2         .        5
3         .        4
4         .        5
4         .        7
```

```
4        .        9
5        .        6
6        .        11
6        .        12
6        .        13
7        .        8
7        .        9
9        .        10
9        .        14
10       .        11
12       .        13
13       .        14/;
conex(bus,node)$(conex(node,bus))=1;
Variables
OF, OBI;
Binary variable PMU(bus),alpha(bus), beta(bus);
Equations const1,const2,const3,const2E,const3A;
const1       .. OF=e=sum(bus,PMU(bus));
const2(bus)  .. PMU(bus)+sum(node$conex(bus,node),PMU
(node))=g=1;
const2E(bus) .. PMU(bus)+sum(node$conex(bus,node),PMU
(node))=g=beta(bus);
const3       .. OBI=e=sum(bus, beta(bus));
const3A      .. OF=e=sum(bus, PMU(bus));
Option optcr=0;
Model placement        /const1,const2/;
Model placement8       /const2E,const3,const3A/;
Set cp /cp1*cp3/; Alias(cp,cpp);
Parameter phase_rep(cp,*),phase_pmu(cp,bus); Parameter
    phase(cp)
/cp1   2
 cp2   1
 cp3   1/;
Solve placement minimizing OF using mip;
display pmu.l; Pmu.fx(bus)$(Pmu.l(bus)=0)=0;
beta.up(bus)=1; beta.lo(bus)=0;
loop(cp,
OF.up=sum(cpp$(ord(cpp)<=ord(cp)),phase(cpp));
solve placement8 maximizing OBI using mip;
phase_rep(cp,'OBS')=OBI.l;
phase_rep(cp,'Tpmu')=sum(bus,Pmu.l(bus));
phase_pmu(cp,bus)=pmu.l(bus);
Pmu.fx(bus)$(PMU.l(bus))=1; );
```

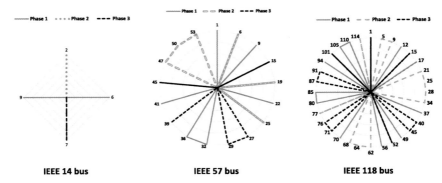

Fig. 8.14 Multistage optimal PMU allocation for different case studies

Table 8.5 Optimal allocation of PMU devices in multistage PMU allocation

Phase	IEEE 14			IEEE 57			IEEE 118		
	OBI	TPMU	ψ_p	OBI	TPMU	ψ_p	OBI	TPMU	ψ_p
I	10	2	2	28	6	6	70	11	11
II	13	3	1	47	12	6	103	22	11
III	14	4	1	57	17	5	118	32	10

The solution for multistage optimal PMU allocation for different case studies is shown in Fig. 8.14.

The optimal allocation of PMU devices in multistage PMU allocation is shown in Table 8.5. It is assumed that the total number of PMU (TPMU) for full observability of the system is known in advance. Three installation phases are considered and the possible number of installable PMU devices (ψ_p) is also known.

8.7 Applications

Different objective functions as well as methods have been proposed in the literature to improve the power system observability as follows:

- MIP formulation for optimal PMU placement [3]
- Multi-stage MIP-based PMU allocation [2]
- Optimal PMU Placement considering measurement Loss and branch outage [4]
- Weighted least squares algorithm for optimal PMU placement [5]
- Redundancy and observability analysis of conventional measurement and PMU [6]
- Semi-definite Programming for optimal PMU placement considering fixed channel capacity [7]

- Optimal PMU placement for full network observability using tabu search algorithm [8]
- Multi-objective optimal PMU placement using a non-dominated sorting differential evolution algorithm [9]
- Iterated Local Search for optimal PMU allocation [10]

References

1. V. Basetti, A.K. Chandel, Optimal PMU placement for power system observability using Taguchi binary bat algorithm. Measurement **95**, 8–20 (2017)
2. D. Dua, S. Dambhare, R.K. Gajbhiye, S.A. Soman, Optimal multistage scheduling of PMU placement: an ILP approach. IEEE Trans. Power Delivery **23**(4), 1812–1820 (2008)
3. B. Gou, Generalized integer linear programming formulation for optimal PMU placement. IEEE Trans. Power Syst. **23**(3), 1099–1104 (2008)
4. C. Rakpenthai, S. Premrudeepreechacharn, S. Uatrongjit, N.R. Watson, An optimal PMU placement method against measurement loss and branch outage. IEEE Trans. Power Delivery **22**(1), 101–107 (2007)
5. N.M. Manousakis, G.N. Korres, A weighted least squares algorithm for optimal PMU placement. IEEE Trans. Power Syst. **28**(3), 3499–3500 (2013)
6. J.B.A. London, S.A.R. Piereti, R.A.S. Benedito, N.G. Bretas, Redundancy and observability analysis of conventional and PMU measurements. IEEE Trans. Power Syst. **24**(3), 1629–1630 (2009)
7. N.M. Manousakis, G.N. Korres, Optimal PMU placement for numerical observability considering fixed channel capacity; a semidefinite programming approach. IEEE Trans. Power Syst. **31**(4), 3328–3329 (2016)
8. J. Peng, Y. Sun, H.F. Wang, Optimal PMU placement for full network observability using Tabu search algorithm. Int. J. Electr. Power Energy Syst. **28**(4), 223–231 (2006)
9. C. Peng, H. Sun, J. Guo, Multi-objective optimal PMU placement using a non-dominated sorting differential evolution algorithm. Int. J. Electr. Power Energy Syst. **32**(8), 886–892 (2010)
10. M. Hurtgen, J.-C. Maun, Optimal PMU placement using iterated local search. Int. J. Electr. Power Energy Syst. **32**(8), 857–860 (2010)

Chapter 9
Topics in Transmission Operation and Planning

This chapter provides a solution for some transmission network operation and planning studies in GAMS. The transmission investment regarding building new lines and power flow controllers (phase shifter), sensitivity factors, and transmission switching have been discussed in this chapter. The GAMS code for solving each optimization problem is developed and discussed.

9.1 Transmission Network Planning

The question to be answered in transmission expansion planning (TEP) is when and which right of way should be selected to build a new line, perform reconductoring, build a new substation, or install power flow controllers. As a matter of fact, obtaining the public acceptance for building new transmission lines has become a challenging issue for the transmission asset owners. This also makes it difficult for transmission system operator to keep the technical and economic performance of transmission network high. The objective function is usually defined as the total operation and planning costs. Different models have been proposed for TEP purpose such as:

- Probabilistic TEP considering load and wind power generation uncertainties [1]
- MIP-based multi-stage TEP considering losses, generator costs, and the $N-1$ security constraints [2]
- Genetic algorithm-based TEP considering demand uncertainty [3]
- Congestion reduction-based TEP [4]
- Robust optimization-based TEP considering the uncertainties of renewable generation and load [5]
- Branch and bound algorithm for TEP [6]
- MIP-based TEP model considering different demand levels, $N-1$ network security constraints as well as environmental constraints [7]

© Springer International Publishing AG 2017
A. Soroudi, *Power System Optimization Modeling in GAMS*,
DOI 10.1007/978-3-319-62350-4_9

- Multi-objective TEP considering total social costs, maximum regret (robustness criterion), and maximum adjustment cost (flexibility criterion) as three objective functions [8]
- Monte-Carlo-based TEP considering random outages of generating units and transmission lines as well as inaccuracies in the long-term load forecasting [9]
- An interior point method considering full AC power flow constraints [10]
- Multi-objective TEP considering investment cost, reliability (both adequacy and security), and congestion costs [11]
- Chance constrained TEP consideration of load and wind uncertainties [12].

9.1.1   TEP with New Lines Option

The transmission expansion planning is formulated in (9.1). It is assumed that the only planning option is building new lines.

$$OF = T \times OPC + INVC \tag{9.1a}$$

$$OPC = \sum_{g \in \Omega_G} b_g P_g + VOLL \sum_i LS_i \tag{9.1b}$$

$$INVC = \left(-\eta_{ij}^0 + \sum_{k,ij} \alpha_{ij}^k \right) C_{ij} \tag{9.1c}$$

$$P_{ij}^k - B_{ij}(\delta_i - \delta_j) \leq (1 - \alpha_{ij}^k)M \tag{9.1d}$$

$$P_{ij}^k - B_{ij}(\delta_i - \delta_j) \geq -(1 - \alpha_{ij}^k)M \tag{9.1e}$$

$$\sum_{g \in \Omega_G^i} P_g + LS_i - L_i = \sum_{j \in \Omega_\ell^i} P_{ij} : \lambda_i \quad i \in \Omega_B \tag{9.1f}$$

$$-P_{ij}^{\max} \alpha_{ij}^k \leq P_{ij}^k \leq P_{ij}^{\max} \alpha_{ij}^k \quad ij \in \Omega_\ell \tag{9.1g}$$

$$P_g^{\min} \leq P_g \leq P_g^{\max} \tag{9.1h}$$

$$\text{if } \eta_{ij}^0 = 1 \text{ then } \alpha_{ij}^{k=1} = 1 \tag{9.1i}$$

$$B_{ij} = \frac{1}{x_{ij}} \tag{9.1j}$$

$$\alpha_{ij}^k \in \{0, 1\} \tag{9.1k}$$

$$k \in \{1, 2, 3, 4\} \tag{9.1l}$$

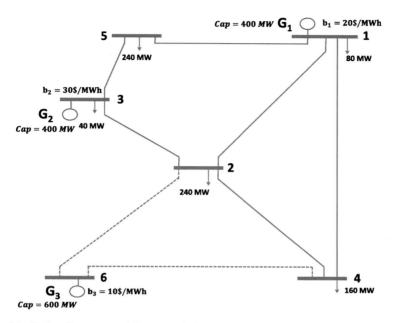

Fig. 9.1 Six-bus Garver transmission network (base case)

The objective function in (9.1a) consists of operational costs (OPC) and investment costs (INVC). The operational costs are calculated in (9.1b) while the investment costs are calculated in (9.1c). (9.1d), and (9.1e) model, the power flow on branch connecting bus i to bus j. In (9.1d) and (9.1e) there is a parameter M. It is also called big M in the literature [13]. This parameter is selected as follows:

$$M = \max_{ij} B_{ij}(\delta_i - \delta_j)$$

The power balance between the generated power, load shedding, demand, and line flows is ensured by (9.1f). The line flow limits are modeled in (9.1g) and the impacts of line investment decision α_{ij}^k are formulated. The generation operating limits are given in (9.1h). The initial status of each line is described in (9.1i).

The six-bus Garver transmission network [14] is shown in Fig. 9.1.

Transmission expansion planning data are given in Table 9.1. The existing and potential right of ways in addition to the reactances and flow limits are provided there. The investment costs (C_{ij}) are given in million \$. The VOLL is assumed to be 1000 \$/MW h. There are three generating units with different operating costs as shown in Fig. 9.1. Although the generator number 3 is the cheapest unit, however, it is not connected to the grid. The candidate right of ways are the existing ones (indicated by solid lines and the dashed ones as indicated in Fig. 9.1). The GAMS code for solving the DC power flow-based TEP is provided in GCode 9.1.

Table 9.1 Transmission expansion planning data for Garver six-bus transmission network

From (i)	To (j)	x_{ij}	$\bar{f}_{ij}$	C_{ij}	η_{ij}^0
1	2	0.40	100	40	1
1	4	0.60	80	60	1
1	5	0.20	100	20	1
2	3	0.20	100	20	1
2	4	0.40	100	40	1
2	6	0.30	100	30	0
3	5	0.20	100	20	1
4	6	0.30	100	30	0

GCode 9.1 DC-OPF-based TEP

```
Sets bus    /1*6/, slack(bus) /1/, Gen /g1*g3/, k /k1*k4/;
Scalars  Sbase /100/ , M /1000/; alias(bus,node);
Table GenData(Gen,*)   Generating  units  characteristics
     b       pmin pmax
g1  20     0     400
g2  30     0     400
g3  10     0     600;
set  GBconect(bus,Gen) connectivity index of each generating unit to each bus
/1       .      g1
 3       .      g2
 6       .      g3   /  ;
Table  BusData(bus,*) Demands  of each  bus  in MW
         Pd
1        80
2        240
3        40
4        160
5        240;
table branch(bus, node,*)      Network technical characteristics
                 X     LIMIT Cost stat
1     .    2    0.4   100   40    1
1     .    4    0.6   80    60    1
1     .    5    0.2   100   20    1
2     .    3    0.2   100   20    1
2     .    4    0.4   100   40    1
2     .    6    0.3   100   30    0
3     .    5    0.2   100   20    1
4     .    6    0.3   100   30    0;
Set  conex(bus,node)           Bus connectivity matrix;
conex(bus,node)$(branch(bus,node,'x'))=yes; conex(bus,node)$conex(node,bus)=yes;
branch(bus,node,'x')$branch(node,bus,'x')=branch(node,bus,'x');
branch(bus,node,'cost')$branch(node,bus,'cost')=branch(node,bus,'cost');
branch(bus,node,'stat')$branch(node,bus,'stat')=branch(node,bus,'stat');
branch(bus,node,'Limit')$(branch(bus,node,'Limit')=0)=branch(node,bus,'Limit');
branch(bus,node,'bij')$conex(bus,node) =1/branch(bus,node,'x');
M=smax((bus,node)$conex(bus,node),branch(bus,node,'bij')*pi*4/3);
Variables   OF,Pij(bus,node,k),Pg(Gen),delta(bus),LS(bus);
binary variable alpha(bus,node,k); alpha.l(bus,node,k)=1;
alpha.fx(bus,node,k)$(conex(bus,node) and ord(k)=1 and  branch(node,bus,'stat') )
     =1;
```

```
Equations const1A , const1B , const1C , const1D , const1E , const2 , const3 ;
const1A ( bus , node , k ) $conex ( node , bus )  .. Pij ( bus , node , k )—
branch ( bus , node , ' bij ' ) *( delta ( bus )—delta ( node ))=l= M*(1—alpha ( bus , node , k ));
const1B ( bus , node , k ) $conex ( node , bus )  .. Pij ( bus , node , k )—
branch ( bus , node , ' bij ' ) *( delta ( bus )—delta ( node ))=g=—M*(1—alpha ( bus , node , k ));
const1C ( bus , node , k ) $conex ( node , bus )  .. Pij ( bus , node , k )=l=
 alpha ( bus , node , k )* branch ( bus , node , ' Limit ' ) / Sbase ;
const1D ( bus , node , k ) $conex ( node , bus )  .. Pij ( bus , node , k )=g=
—alpha ( bus , node , k )* branch ( bus , node , ' Limit ' ) / Sbase ;
const1E ( bus , node , k ) $conex ( node , bus )  .. alpha ( bus , node , k )=e=alpha ( node , bus , k );
const2 ( bus )  .. LS ( bus )+sum ( Gen$GBconect ( bus , Gen ) , Pg ( Gen ) )—BusData ( bus , ' pd ' ) / Sbase
=e=+sum (( k , node ) $conex ( node , bus ) , Pij ( bus , node , k ));
const3         .. OF=g=2*8760*(sum ( Gen , Pg ( Gen )* GenData ( Gen , ' b ' )* Sbase )
+1000* sbase*sum ( bus , LS ( bus ))  )
+1e6*sum (( bus , node , k ) $conex ( node , bus ) ,0.5* branch ( bus , node , ' cost ' )* alpha ( bus , node , k )
$( ord ( k )>1  or  branch ( node , bus , ' stat ' )=0));
Model  loadflow        / all /; option  optcr =0;
LS . up ( bus )=BusData ( bus , ' pd ' ) / Sbase ;  LS . lo ( bus )=0;
Pg . lo ( Gen )=GenData ( Gen , ' Pmin ' ) / Sbase ;
Pg . up ( Gen )=GenData ( Gen , ' Pmax ' ) / Sbase ;
delta . up ( bus )=pi /3;  delta . lo ( bus )=—pi /3;  delta . fx ( slack )=0;
Pij . up ( bus , node , k ) $(( conex ( bus , node )))=1* branch ( bus , node , ' Limit ' ) / Sbase ;
Pij . lo ( bus , node , k ) $(( conex ( bus , node )))=—1* branch ( bus , node , ' Limit ' ) / Sbase ;
Solve  loadflow  min  OF  us  MIP ;
```

9.1.1.1 Two Years Planning Period

The optimal TEP solution for 2 years planning period ($T = 2 * 8760$ h) is shown in Fig. 9.2. The total operating cost is M\$227.142, and the investment costs are M\$110. The total costs would be M\$337.140.

The optimal solutions dictate that the branch connecting bus 1 to bus 5 should be reinforced with one additional line. Bus 6 to bus 2 should be connected using three lines.

9.1.1.2 Ten Years Planning Period

The optimal TEP solution for 2 years planning period ($T = 10 * 8760$ h) is shown in Fig. 9.3. The total operating cost is M\$871.410 and the investment costs are M\$162.853. The total costs would be M\$1034.263. The optimal solutions dictates that the branch$_{2-3}$ and branch$_{1-5}$ should be reinforced with one additional line. Bus 6 to bus 2 should be connected using four lines. Bus 6 to bus 4 should be connected using two lines. It should be noted that the current model is simplistic. It needs to consider more realistic constraints. Some of them are listed as follows:

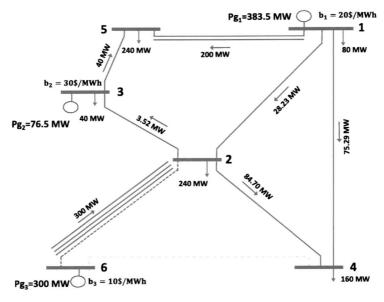

Fig. 9.2 Optimal TEP solution for 2 years planning period ($T = 2 * 8760$ h)

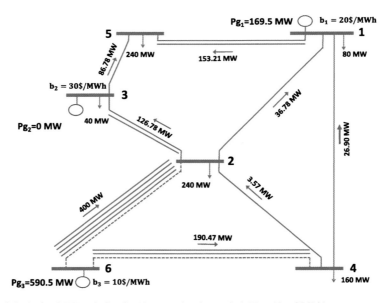

Fig. 9.3 Optimal TEP solution for 10 years planning period ($T = 10 * 8760$ h)

- The model is a single period. The load duration curve (LDC) or a discrete LDC with some demand levels and their associated duration should be considered.
- The contingencies should be taken into account. This will ensure the system operator to keep the light on even some network/generation assets fail. The failure of transmission lines or generating units might cause overloading of the remaining lines and operation of over-current relays and cause cascaded failures. It might even lead to black out.
- The AC power flow constraints should be checked to make sure no voltage constraint or line limit is violated.
- The voltage stability issue should also be checked.
- The model is trying to minimize the total operating cost plus the investment costs. In deregulated environment, the transmission asset owner and the transmission system operator are two independent entities. This makes the problem more complicated since these two entities would have different objective functions. The asset owner is trying to maximize its benefit and make money by making investments. On the other hand, the system operator is trying to maximize the social welfare. The multi-objective techniques [8] or complementarity models [15] provide the suitable answer to this challenge.
- The demand grows should be considered.
- The formulated transmission planning model is a static model. This means that the decision is made in order to make the system capable of answering for the needs of next N years. The dynamic of investment or timing of investment and time value of money is neglected.
- The uncertainties of the electricity market, renewable energies, demand, and regulatory frameworks should be considered to make the model robust against future scenarios.
- The VOLL is assumed to be the same for all demands at different buses. However, the importance level of all demands is not the same. This can reduce the investment requirements for TEP.

9.1.2 TEP with New Lines and Power Flow Controller Option

This section investigates the impact of power flow controller as a planning option. The phase shifter is used as the power flow controller device. The role of the phase shifter is depicted in Fig. 9.4. The relation between the power flow, voltage angles difference across the branch$_{ij}$, and the line susceptance is given in (9.2).

$$P_{ij} = \frac{\delta_i - \delta_j + \Psi_{ij}}{x_{ij}} = B_{ij}(\delta_i - \delta_j + \Psi_{ij}) \tag{9.2a}$$

$$-\Psi_{ij}^{max} \leq \Psi_{ij} \leq +\Psi_{ij}^{max} \tag{9.2b}$$

$$\Psi_{ij} = -\Psi_{ji} \tag{9.2c}$$

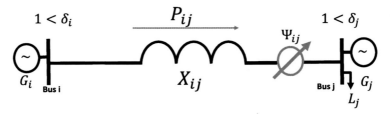

Fig. 9.4 The phase shifter function in power flow control

Although the voltage phase shift is a discrete variable in reality but it is modeled as a continuous variable for simplicity in (9.2).

$$\text{OF} = T \times \text{OPC} + \text{INVC} \tag{9.3a}$$

$$\text{OPC} = \sum_{g \in \Omega_G} b_g P_g + \text{VOLL} \sum_i \text{LS}_i \tag{9.3b}$$

$$\text{INVC} = \left(-\eta_{ij}^0 + \sum_{k,ij} \alpha_{ij}^k \right) C_{ij} + \sum_{k,ij} I_{ij}^k \gamma_{ij} \tag{9.3c}$$

$$P_{ij}^k - B_{ij}(\delta_i - \delta_j + \Psi_{ij}^k) \le (1 - \alpha_{ij}^k)M \tag{9.3d}$$

$$P_{ij}^k - B_{ij}(\delta_i - \delta_j + \Psi_{ij}^k) \ge -(1 - \alpha_{ij}^k)M \tag{9.3e}$$

$$-\Psi_{ij}^{\max} I_{ij}^k \le \Psi_{ij}^k \le +\Psi_{ij}^{\max} I_{ij}^k \tag{9.3f}$$

$$\Psi_{ij} = -\Psi_{ji} \tag{9.3g}$$

$$I_{ij}^k \le \alpha_{ij}^k \tag{9.3h}$$

$$\sum_{g \in \Omega_G^i} P_g + \text{LS}_i - L_i = \sum_{j \in \Omega_\ell^i} P_{ij} : \lambda_i \quad i \in \Omega_B \tag{9.3i}$$

$$-P_{ij}^{\max} \alpha_{ij}^k \le P_{ij}^k \le P_{ij}^{\max} \alpha_{ij}^k \quad ij \in \Omega_\ell \tag{9.3j}$$

$$P_g^{\min} \le P_g \le P_g^{\max} \tag{9.3k}$$

$$\text{if } \eta_{ij}^0 = 1 \text{ then } \alpha_{ij}^{k=1} = 1 \tag{9.3l}$$

$$B_{ij} = \frac{1}{x_{ij}} \tag{9.3m}$$

$$\alpha_{ij}^k, I_{ij}^k \in \{0, 1\} \tag{9.3n}$$

$$k \in \{1, 2, 3, 4\} \tag{9.3o}$$

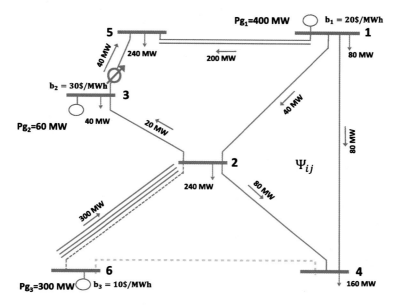

Fig. 9.5 The optimal decisions regarding the location of phase shifter and new transmission lines 2 years plan ($T = 2 * 8760$)

The phase shifter impact is shown in (9.3d) and (9.3e). The investment decision regarding the phase shifter is reflected in (9.3f). The phase shifter can be installed on lines that initially exist or built lines as shown in (9.3h). The investment cost for each phase shifter is assumed to be $\gamma_{ij} = $ M\$0.6. The optimal decisions regarding the location of phase shifter and new transmission lines 2 years plan ($T = 2 * 8760$) are given in Fig. 9.5. The total operating cost is M\$224.256 and the investment costs are M\$110.6. The total costs would be M\$334.856.

GCode 9.2 DC-OPF-based TEP considering phase shifter option

```
Sets bus   /1*6/ ,slack(bus) /1/,Gen /g1*g3/,k /k1*k4/; alias(bus,node);
Scalars   Sbase /100/ , M /1000/, T; T=8760*2; Set conex(bus,node);
Table GenData(Gen,*)   Generating units characteristics
     b      pmin pmax
g1  20      0    400
g2  30      0    400
g3  10      0    600;
Set GBconect(bus,Gen) connectivity index of each generating unit to each bus
/1    .    g1
 3    .    g2
 6    .    g3  / ;
Table BusData(bus,*) Demands of each bus in MW
          Pd
1         80
2         240
```

```
3         40
4        160
5        240;
table branch(bus, node,*)      Network  technical  characteristics
                 X     LIMIT  Cost  stat
;
conex(bus,node)$(branch(bus,node,'x'))=yes;  conex(bus,node)$conex(node,bus)=yes;
branch(bus,node,'x')$branch(node,bus,'x')=branch(node,bus,'x');
branch(bus,node,'cost')$branch(node,bus,'cost')=branch(node,bus,'cost');
branch(bus,node,'stat')$branch(node,bus,'stat')=branch(node,bus,'stat');
branch(bus,node,'Limit')$(branch(bus,node,'Limit')=0)=branch(node,bus,'Limit');
branch(bus,node,'bij')$conex(bus,node) =1/branch(bus,node,'x');
M=smax((bus,node)$conex(bus,node),branch(bus,node,'bij')*3.14*2);
Variables   OF, Pij(bus,node,k),Pg(Gen),delta(bus),LS(bus),PSHij(bus,node,k);
binary  variable  alpha(bus,node,k),I(bus,node,k);  alpha.l(bus,node,k)=1;
alpha.fx(bus,node,k)$(conex(bus,node)  and  ord(k)=1  and  branch(node,bus,'stat') )
   =1;
Equations  const1A ,const1B ,const1C ,const1D ,const1E ,const2 ,const3 ,
const4 ,const5 ,const6 ,const7 ;
const1A(bus,node,k)$conex(node,bus)  ..  Pij(bus,node,k)-branch(bus,node,'bij')*(
               delta(bus)-delta(node)+PSHij(bus,node,k))=l=M*(1-alpha(bus,node,k))
                ;
const1B(bus,node,k)$conex(node,bus)  ..  Pij(bus,node,k)-branch(bus,node,'bij')*(
               delta(bus)-delta(node)+PSHij(bus,node,k))=g=-M*(1-alpha(bus,node,k))
                ;
const1C(bus,node,k)$conex(node,bus)  ..  Pij(bus,node,k)=l=
alpha(bus,node,k)*branch(bus,node,'Limit')/Sbase;
const1D(bus,node,k)$conex(node,bus)  ..  Pij(bus,node,k)=g=
-alpha(bus,node,k)*branch(bus,node,'Limit')/Sbase;
const1E(bus,node,k)$conex(node,bus)  ..  alpha(bus,node,k)=e=alpha(node,bus,k);
const2(bus)  ..  LS(bus)+sum(Gen$GBconect(bus,Gen),Pg(Gen))
-BusData(bus,'pd')/Sbase=e=+sum((k,node)$conex(node,bus),Pij(bus,node,k));
const3  ..OF=g=T*(sum(Gen,Pg(Gen)*GenData(Gen,'b')*Sbase)+100000*sum(bus,LS(bus)))
+1e6*sum((bus,node,k)$conex(node,bus),0.5*branch(bus,node,'cost')*
alpha(bus,node,k)$(ord(k)>1 or branch(node,bus,'stat')=0))+6e5*0.5*sum((bus,node,k)
$conex(node,bus),I(bus,node,k));
const4(bus,node,k)$conex(node,bus)  ..PSHij(bus,node,k)+PSHij(node,bus,k)=e=0;
const5(bus,node,k)$conex(node,bus)  ..PSHij(bus,node,k)=l=I(bus,node,k)*pi/8;
const6(bus,node,k)$conex(node,bus)  ..PSHij(bus,node,k)=g=-I(bus,node,k)*pi/8;
const7(bus,node,k)$conex(node,bus)  ..I(bus,node,k)=l=alpha(bus,node,k);
Model  loadflow     / all /;  LS.up(bus)=BusData(bus,'pd')/Sbase;  LS.lo(bus)=0;
Pg.lo(Gen)=GenData(Gen,'Pmin')/Sbase;  Pg.up(Gen)=GenData(Gen,'Pmax')/Sbase;
delta.up(bus)=pi/3;  delta.lo(bus)=-pi/3;  delta.fx(slack)=0;
Pij.up(bus,node,k)$((conex(bus,node)))=1*branch(bus,node,'Limit')/Sbase;
Pij.lo(bus,node,k)$((conex(bus,node)))=-1*branch(bus,node,'Limit')/Sbase;
PSHij.up(bus,node,k)=  pi/8;  PSHij.lo(bus,node,k)=-pi/8;  option optcr=0;
Solve  loadflow  min  OF  us  MIP;
```

The optimal decisions regarding the location of phase shifter and new transmission lines 10 years plan ($T = 10 * 8760$) are given in Fig. 9.6. The total operating cost is M\$805.920 and the investment costs are M\$200.600. The total costs would be M\$1006.520.

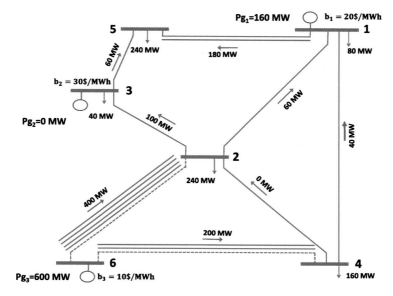

Fig. 9.6 The optimal decisions regarding the location of phase shifter and new transmission lines 10 years plan ($T = 10 * 8760$)

Nomenclature

Indices and Sets

g	Index of thermal generating units
i, j	Index of network buses
Ω_G	Set of all thermal generating units
Ω_G^i	Set of all thermal generating units connected to bus i
Ω_ℓ^i	Set of all buses connected to bus i
Ω_B	Set of network buses

Parameters

M	Big number
T	Duration of planning period (h)
L_i	Electric power demand in bus i at time t
b_g	Fuel cost coefficient of thermal unit g
η_{ij}^0	Initial status of branch connecting bus i to j
C_{ij}	Investment cost for branch connecting bus i to j
γ_{ij}	Investment cost of phase shifter in line ij
$P_g^{\max / \min}$	Maximum/minimum limits of power generation of thermal unit g

P_{ij}^{max} Maximum power flow limits of branch connecting bus i to j
Ψ_{ij}^{max} Maximum phase shift in line connecting bus i to bus j
x_{ij} Reactance of branch connecting bus i to j
B_{ij} Susceptance of branch connecting bus i to j
VOLL   Value of loss of load (\$/MW h)

Variables

P_{ij}^k Active power flow of branch k connecting bus i to j (MW)
P_g Active power generated by thermal unit g (MW)
α_{ij}^k Binary variable to model the investment decision regarding the line k at the right of way ij
I_{ij}^k Binary variable to model the investment decision regarding the phase shifter in line k at the right of way ij
λ_i Locational marginal price in bus i (\$/MW h)
LS$_i$ Load shedding in bus i (MW)
Ψ_{ij} Phase shift in line connecting bus i to bus j
OPC   Total operating costs (\$)
OF   Total costs (\$)
INVC   Total investment costs (\$)
δ_i Voltage angle of bus i (rad)

9.2   Sensitivity Factors in Transmission Networks

In this section two important factors are analyzed and calculated namely:

• Generation Shift Factor (GSF)
• Line Outage Distribution Factor (LODF)

9.2.1   Generation Shift Factors

The transmission network planner/operator is always interested to know what happens to the line flows if any outage happens in generation units. In other words, what is the influence of generation/demand change in bus i on the line ℓ (connecting bus n to bus m)? It is important since there is always a chance for contingencies to happen in generating units. This section will provide an answer to this question.

The DC power flow equations of the network shown in Fig. 9.7 is provided in (9.4).

$$P_1^g - L_1 = \frac{\delta_1 - \delta_2}{x_{12}} + \frac{\delta_1 - \delta_3}{x_{13}} \tag{9.4a}$$

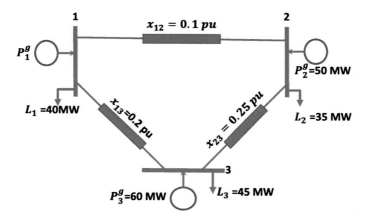

Fig. 9.7   Three-bus network example for sensitivity factors calculation

$$P_2^g - L_2 = \frac{\delta_2 - \delta_1}{x_{21}} + \frac{\delta_2 - \delta_3}{x_{23}} \qquad (9.4b)$$

$$P_3^g - L_3 = \frac{\delta_3 - \delta_2}{x_{32}} + \frac{\delta_3 - \delta_1}{x_{31}} \qquad (9.4c)$$

By substituting the numerical values of x_{ij} in (9.4) we will have :

$$P_1^g - L_1 = \frac{\delta_1 - \delta_2}{0.1} + \frac{\delta_1 - \delta_3}{0.2} \qquad (9.5a)$$

$$P_2^g - L_2 = \frac{\delta_2 - \delta_1}{0.1} + \frac{\delta_2 - \delta_3}{0.25} \qquad (9.5b)$$

$$P_3^g - L_3 = \frac{\delta_3 - \delta_2}{0.25} + \frac{\delta_3 - \delta_1}{0.2} \qquad (9.5c)$$

(9.4) can be written as : $\begin{pmatrix} P_1 \\ P_2 \\ P_3 \end{pmatrix} = \begin{pmatrix} 15 & -10 & -5 \\ -10 & 14 & -4 \\ -5 & -4 & 9 \end{pmatrix} = \begin{pmatrix} \delta_1 \\ \delta_2 \\ \delta_3 \end{pmatrix}$ where $P_i = P_i^g - L_i$.

Suppose $B = \begin{pmatrix} 15 & -10 & -5 \\ -10 & 14 & -4 \\ -5 & -4 & 9 \end{pmatrix}$ then we can have the relation between the bus

angles and the net injections as a linear matrix form.

$$P = B\delta \qquad (9.6)$$

In normal DC power flow, the P vector is known and the decision maker's goal is to find the δ vector. The problem is that matrix B is singular and does not have a matrix inverse.

The good news is that the bus angle at slack bus is known to be zero. If the slack bus is bus 1 then we can have the following matrix form:

$$\begin{pmatrix} P_2 \\ P_3 \end{pmatrix} = \begin{pmatrix} 14 & -4 \\ -4 & 9 \end{pmatrix} \begin{pmatrix} \delta_2 \\ \delta_3 \end{pmatrix} \tag{9.7}$$

Now the matrix can be inversed as follows

$$\begin{pmatrix} \delta_2 \\ \delta_3 \end{pmatrix} = \begin{pmatrix} 14 & -4 \\ -4 & 9 \end{pmatrix}^{-1} \begin{pmatrix} P_2 \\ P_3 \end{pmatrix} = \begin{pmatrix} 0.0818 & 0.0364 \\ 0.0364 & 0.1273 \end{pmatrix} \begin{pmatrix} P_2 \\ P_3 \end{pmatrix} \tag{9.8}$$

We can write down (9.8) as a general form

$$\delta_{\text{red}} = X_{\text{red}} P_{\text{red}} \tag{9.9}$$

$$X_{\text{red}} = B_{\text{red}}^{-1} \tag{9.10}$$

where B_{red} is the B matrix after eliminating the row and column of slack bus. If the network has n buses then B_{red} would be a square $(n-1) \times (n-1)$ matrix. For the rest of this chapter, whenever X appears in any equation it is representing X_{red}.

Now we are about to investigate the impact of change in power injection in bus n on the line flow between bus i and bus j. The flow on line connecting bus i to bus j is calculated as follows:

$$f_{ij} = \frac{\delta_i - \delta_j}{x_{ij}} \tag{9.11}$$

Now we assume that any change in injected power at bus m will be compensated by the slack bus. In order to calculate the flow change in line $i - j$ we need to use the following equation:

$$\Delta f_{ij} = \frac{\Delta \delta_i - \Delta \delta_j}{x_{ij}} \tag{9.12}$$

The value of x_{ij} remains constant but the voltage angles would change if ΔP is happening at bus m (and $-\Delta P$ at slack bus). Referring to (9.9) we would have

$$\Delta \delta = X \Delta P \tag{9.13}$$

$$
\begin{pmatrix} \Delta\delta_2 \\ \Delta\delta_3 \\ \vdots \\ \Delta\delta_m \\ \vdots \\ \Delta\delta_i \\ \vdots \\ \Delta\delta_j \\ \vdots \\ \Delta\delta_{n-1} \\ \Delta\delta_n \end{pmatrix}_{n-1\times1}
=
\begin{pmatrix}
X_{22} & \cdots & X_{2,n-1} & X_{2n} \\
X_{32} & \cdots & X_{3,n-1} & X_{3n} \\
\vdots & \ddots & \ddots & \vdots \\
X_{m2} & \cdots & X_{m,n-1} & X_{mn} \\
\vdots & \ddots & \ddots & \vdots \\
X_{i2} & \cdots & X_{i,n-1} & X_{in} \\
\vdots & \ddots & \ddots & \vdots \\
X_{j2} & \cdots & X_{j,n-1} & X_{jn} \\
\vdots & \ddots & \ddots & \vdots \\
X_{n-1,2} & \cdots & X_{n-1,n-1} & X_{n-1,n} \\
X_{n,2} & \cdots & X_{n,n-1} & X_{n,n}
\end{pmatrix}_{n-1\times n-1}
\begin{pmatrix} 0 \\ 0 \\ \vdots \\ \Delta P_m \\ \vdots \\ 0 \\ \vdots \\ 0 \\ \vdots \\ 0 \\ 0 \end{pmatrix}_{n-1\times1}
\tag{9.14}
$$

The changes of voltage angles in bus i and j are calculated as follows:

$$\Delta\delta_i = X_{im}\Delta P_m \tag{9.15}$$

$$\Delta\delta_j = X_{jm}\Delta P_m \tag{9.16}$$

$$\Delta f_{ij} = \frac{X_{im}\Delta P_m - X_{jm}\Delta P_m}{x_{ij}} \tag{9.17}$$

The (9.17) states that the sensitivity of the flow in line ij of power change in bus m is obtained as follows:

$$a_m^{ij} = \frac{\Delta f_{ij}}{\Delta P_m} = \frac{X_{im} - X_{jm}}{x_{ij}} \tag{9.18}$$

Using the (9.18) is useful in calculating the line flows in post-contingency period.

$$f_{ij}^{\text{post}} = f_{ij}^{\text{pre}} + a_m^{ij}\Delta P_m \tag{9.19}$$

If the post-contingency line flow at line ij after the failure of a generating unit at bus m (producing P_m^g MW) is to be calculated then the following equation can be used:

$$f_{ij}^{\text{post}} = f_{ij}^{\text{pre}} + a_m^{ij}(-P_m^g) \tag{9.20}$$

If the post-contingency line flow at line ij after the disconnection of load at bus m is to be calculated then the following equation can be used:

$$f_{ij}^{\text{post}} = f_{ij}^{\text{pre}} + a_m^{ij}(L_m) \tag{9.21}$$

Now let's calculate the Generation Shift Factors (a_m^{ij}) for the network shown in Fig. 9.7.

The GCode 9.3 described the GAMS code for calculating the generation shift factors.

GCode 9.3 Generation Shift Factor Calculation

```
Sets bus  /1*3/ ,slack(bus) /1/,Gen /g1*g3/, nonslack(bus) /2*3/ ;
scalars  Sbase /100/ ;
alias(bus,node,shin,knot); alias(nonslack,nonslackj);
Table branch(bus,node,*)        Network technical characteristics
                 X       LIMIT  stat
1     .    2    0.1     100     1
1     .    3    0.2     80      1
2     .    3    0.25    100     1;
set conex(bus,node)             Bus connectivity matrixl;
conex(bus,node)$(branch(bus,node,'x'))=yes;
conex(bus,node)$conex(node,bus)=yes;
branch(bus,node,'x')$branch(node,bus,'x')=branch(node,bus,'x');
branch(bus,node,'stat')$branch(node,bus,'stat')=branch(node,bus,'stat');
branch(bus,node,'Limit')$(branch(bus,node,'Limit')=0)=branch(node,bus,'
     Limit');
branch(bus,node,'bij')$conex(bus,node) =1/branch(bus,node,'x');
Parameter Bmatrix(bus,node),Binv(bus,node),Flow(bus,node); Alias(bus,
     knot);
Bmatrix(bus,node)$(conex(node,bus))=-branch(bus,node,'bij');
Bmatrix(bus,bus)=sum(knot$conex(knot,bus),-Bmatrix(bus,knot));
parameter Breduced(nonslack,nonslackj),GSHF(bus,node,knot);
Breduced(nonslack,nonslackj)=Bmatrix(nonslack,nonslackj);
parameter inva(nonslack,nonslackj) 'inverse of a';

execute_unload 'a.gdx',nonslack,Breduced;
execute '=invert.exe a.gdx nonslack Breduced b.gdx inva';
execute_load 'b.gdx',inva;

Binv(nonslack,nonslackj)=inva(nonslack,nonslackj);
GSHF(bus,node,knot)$conex(bus,node)=
branch(bus,node,'bij')*(Binv(bus,knot)-Binv(node,knot));

Display Bmatrix,Binv,GSHF;
```

The GCode 9.3 has no solve statement or variable. This is because no optimization is going to be done. First of all, the B matrix is calculated in GCode 9.3 as follows:

```
Parameter Bmatrix(bus,node);
Alias(bus,knot);
Bmatrix(bus,node)$conex(node,bus)=-branch(bus,node,'bij');
Bmatrix(bus,bus)=sum(knot$conex(knot,bus),-Bmatrix(bus,knot));
```

This would calculate the Bmatrix as follows:

```
—— 68 PARAMETER Bmatrix
        1         2         3
 1   15.000   -10.000   -5.000
 2  -10.000    14.000   -4.000
 3   -5.000    -4.000    9.000
```

Now we need to eliminate the row and column containing the slack bus. In order to do this, another parameter called Breduced is defined but over the non-slack bus set. This set is defined over all buses except the slack buses.

```
Parameter Breduced(nonslack,nonslackj);
Breduced(nonslack,nonslackj)=Bmatrix(nonslack,nonslackj);
Parameter inva(nonslack,nonslackj) 'inverse of a';
execute_unload 'a.gdx', nonslack, Breduced;
execute '=invert.exe a.gdx nonslack Breduced b.gdx inva';
execute_load 'b.gdx', inva;
```

This would calculate the inverse matrix of the reduced B matrix.

```
—— 68 PARAMETER Binv
        2        3
 2    0.082    0.036
 3    0.036    0.127
```

Now all needed data for calculating the a_m^{ij} is available. The generation shift factors are calculated as follows:

```
Binv(nonslack,nonslackj)=inva(nonslack,nonslackj);
GSHF(bus,node,knot)$conex(bus,node)=branch(bus,node,'bij')*(Binv
(bus,knot)-Binv(node,knot));
```

Using the GCode 9.3, the generation shift factors are calculated as Table 9.2.

The generation shift factors (GSHF) have the following interesting features as follows:

- The procedure for calculating the GSHF does not involve any optimization
- The values of GSHF can be calculated in advance and be used in real-time applications.
- The values of GSHF do not depend on the loading condition of the network. These coefficients only depend on the network topology. If the network topology is changed (due to transmission outage or switching), GSHF should be recalculated.
- The technique we used for calculating the GSHF is assuming that the changes at any bus are quickly compensated by the slack bus. In case the changes are compensated by multiple generating units then the calculation procedure would be slightly different [16].
- The GSHF are also useful for understanding how to reduce the line loading. For example, suppose we need to reduce the flow at line $3 - 1$. As we can see that bus 3 has the largest GSHF equal to 0.636. It means that if we can reduce the generation at bus 3, then a negative value will be added to the initial flow of line $3 - 1$. If the initial flow from bus 3 to bus 1 is positive, then it would reduce.

Now let's check the values obtained in Table 9.2. For this purpose, the base power flow is solved and is shown in Fig. 9.8. The values of line flow, as well as the voltage angles, are specified in this figure. Suppose it is desired to increase the flow on the line connecting the bus 3 to bus 2. The power flow on this line is $f_{32}^0 = 2.7\,\text{MW}$. Using the Table 9.2 states that $a_2^{32} = -0.182$ and $a_3^{32} = 0.364$. Let's investigate the impact of changes in power injections in different buses on transmission lines.

9.2.1.1 Demand Increase in L_2 by 10 MW

The demand in bus 2 is increased by 10 MW and the new line flows are depicted in Fig. 9.9. As it can be seen in this figure, the new line flow of line $3 - 2$ is 4.5 MW. This is obtained using a GAMS code. Let's calculate the new line flow using the Table 9.2. The new line flow is calculated as follows:

$$f_{32}^{post} = f_{32}^{pre} + a_2^{32}(\Delta P_2) = 0.027 - 0.182 * (-0.1) = 0.0452 \text{ pu} \qquad (9.22)$$

$$f_{12}^{post} = f_{12}^{pre} + a_2^{12}(\Delta P_2) = -0.177 - 0.818 * (-0.1) = -0.0952 \text{ pu} \qquad (9.23)$$

Table 9.2 Generator shift factors (a_m^{ij}) for three-bus network

Line	Bus (m)	
ij	2	3
1–2	−0.818	−0.364
1–3	−0.182	−0.636
2–1	0.818	0.364
2–3	0.182	−0.364
3–1	0.182	0.636
3–2	−0.182	0.364

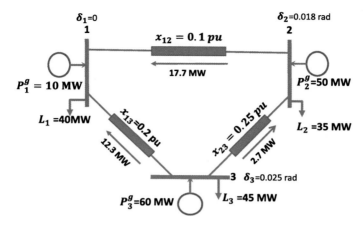

Fig. 9.8 The DC power flow solution for three-bus network

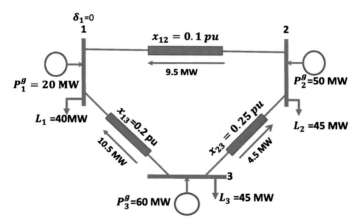

Fig. 9.9 The DC power flow solution for three-bus network after increasing the demand at bus 2 for 10 MW.

$$f_{31}^{post} = f_{31}^{pre} + a_2^{31}(\Delta P_2) = 0.123 + 0.182 * (-0.1) = 0.1048 \text{ pu} \qquad (9.24)$$

It should be noted that the flow values as well as the change in power injection at bus 2 are expressed in pu. ΔP_2 is representing the change in bus injection at bus 2 which is -10 MW or 0.1 pu. It can be observed that the results confirm what is obtained by solving the DC power flow as shown in Fig. 9.9.

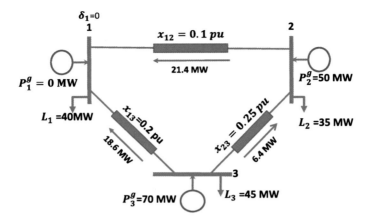

Fig. 9.10 The DC power flow solution for three-bus network after increasing the generation at bus 3 for 10 MW.

9.2.1.2   Generation Increase in P_3^g by 10 MW

The generation in bus 3 is increased by 10 MW and the new line flows are depicted in Fig. 9.10 which are obtained using a GAMS code. Let's calculate the new line flow using the GSHF in Table 9.2. The new line flow is calculated as follows:

$$f_{32}^{post} = f_{32}^{pre} + a_3^{32}(\Delta P_3) = 0.027 + 0.364 * (0.1) = -0.0634 \text{ pu} \qquad (9.25)$$

$$f_{12}^{post} = f_{12}^{pre} + a_3^{12}(\Delta P_3) = -0.177 + 0.364 * (0.1) = -0.2134 \text{ pu} \qquad (9.26)$$

$$f_{31}^{post} = f_{31}^{pre} + a_3^{31}(\Delta P_3) = 0.123 + 0.636 * (0.1) = 0.1866 \text{ pu} \qquad (9.27)$$

9.2.2   Line Outage Distribution Factors

The impact of line outages on power flow of other lines is investigated in this section. Consider the line connecting the bus n and m as shown in Fig. 9.11. In Fig. 9.11a, the intact network is shown. We need to find out the impact of the line outage of the branch connecting bus n to bus m on the rest of the network. Suppose the flow of this line is initially equal to f_{nm}^0. The power from the rest of the network injected to bus n is equal to power absorption from bus m to the rest of network when no contingency has happened. We need to find a way to make these flows equal to zero. It is done using a very smart trick [17]. If we add two injections to the network: $+\Delta P_n$ at bus n and another one equal to $-\Delta P_n$ at bus m the flow on the line nm would change as shown in Fig. 9.11b. The question is what is the new flow on this line? The change of flow on line nm can be easily calculated using the following equation:

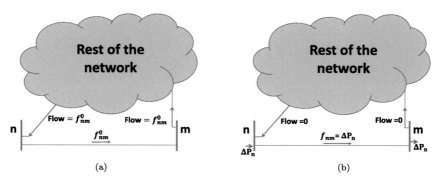

Fig. 9.11 Line outage modeling using virtual injections. (**a**) Intact network. (**b**) Post-contingency network

$$\Delta f_{nm} = \frac{\Delta\delta_n - \Delta\delta_m}{x_{nm}} = \frac{X_{nn}\Delta P_n + X_{nm}(-\Delta P_n) - (X_{mn}\Delta P_n + X_{mm}(-\Delta P_n))}{x_{nm}}$$

(9.28)

This means that

$$\Delta f_{nm} = \frac{X_{nn} + X_{mm} - 2 * X_{nm}}{x_{nm}}\Delta P_n$$

(9.29)

Now the post-contingency line flow is calculated as follows:

$$f_{nm}^{\text{post}} = f_{nm}^0 + \Delta f_{nm}$$

(9.30)

If the virtual injection to the grid ΔP_n is carefully chosen then $f_{nm}^{\text{post}} = \Delta P_n$. This makes the flow from the rest of the network to bus n and m equal to zero (line outage).

$$\Delta P_n = f_{nm}^{\text{post}} = f_{nm}^0 + \Delta f_{nm}$$

(9.31)

Combining the (9.29) with (9.31) gives us :

$$\Delta P_n = f_{nm}^0 + \frac{X_{nn} + X_{mm} - 2X_{nm}}{x_{nm}}\Delta(P_n)$$

(9.32)

$$\Delta P_n\left(1 - \frac{X_{nn} + X_{mm} - 2X_{nm}}{x_{nm}}\right) = f_{nm}^0$$

$$\Delta P_n = \frac{f_{nm}^0}{\left(1 - \frac{X_{nn}+X_{mm}-2X_{nm}}{x_{nm}}\right)}$$

Now the change in power flow in line ij is calculated as follows:

$$\Delta f_{ij} = \frac{\Delta \delta_i - \Delta \delta_j}{x_{ij}} = \frac{X_{in}\Delta P_n + X_{im}(-\Delta P_n) - (X_{jn}\Delta P_n + X_{jm}(-\Delta P_n))}{x_{ij}} \quad (9.33)$$

$$\Delta f_{ij} = \frac{X_{in} - X_{im} - X_{jn} + X_{jm}}{x_{ij}} \Delta P_n \quad (9.34)$$

$$\text{LODF}_{ij,nm} = \frac{\Delta f_{ij}}{f_{nm}^0} = \frac{X_{in} - X_{im} - X_{jn} + X_{jm}}{x_{ij}(1 - \frac{X_{nn}+X_{mm}-2X_{nm}}{x_{nm}})} \quad (9.35)$$

The GCode 9.4 is developed to calculate the Line outage distribution factors for the network shown in Fig. 9.7.

GCode 9.4 Line outage distribution factor calculation

```
Sets
bus    /1*3/ ,
slack(bus) /1/,
Gen /g1*g3/,
nonslack(bus) /2*3/ ;

Scalars  Sbase /100/ ;

Alias(bus,node,shin,knot);
Alias(nonslack,nonslackj);

Table branch(bus,node,*)        Network technical characteristics
                    X    LIMIT  stat
1    .    2    0.1    100    1
1    .    3    0.2    80     1
2    .    3    0.25   100    1;

Set  conex(bus,node)            Bus  connectivity  matrix1;

conex(bus,node)$(branch(bus,node,'x'))=yes;
conex(bus,node)$conex(node,bus)=yes;

branch(bus,node,'x')$branch(node,bus,'x')=branch(node,bus,'x');
branch(bus,node,'stat')$branch(node,bus,'stat')=branch(node,bus,'stat');
branch(bus,node,'Limit')$(branch(bus,node,'Limit')=0)=branch(node,bus,'
    Limit');
branch(bus,node,'bij')$conex(bus,node) =1/branch(bus,node,'x');

Parameter  Bmatrix(bus,node),Binv(bus,node);
Alias(bus,knot);

Bmatrix(bus,node)$(conex(node,bus))=-branch(bus,node,'bij');
Bmatrix(bus,bus)=sum(knot$conex(knot,bus),-Bmatrix(bus,knot));
parameter  Breduced(nonslack,nonslackj),GSHF(bus,node,knot),X0(bus,node);
Breduced(nonslack,nonslackj)=Bmatrix(nonslack,nonslackj);

Parameter  inva(nonslack,nonslackj)  'inverse  of  a',Dfactor(bus,node,knot,
    shin),
```

```
contingency(bus,node,knot,shin);

execute_unload 'a.gdx',nonslack,Breduced;
execute '=invert.exe a.gdx nonslack Breduced b.gdx inva';
execute_load 'b.gdx',inva;

Binv(nonslack,nonslackj)=inva(nonslack,nonslackj);
GSHF(bus,node,knot)$conex(bus,node)=
branch(bus,node,'bij')*(Binv(bus,knot)−Binv(node,knot));

dfactor(bus,node,knot,shin)$( conex(bus,node) and   conex(knot,shin)
and  (ord(bus)>ord(node))
and  (ord(knot)>ord(shin))
and  (ord(bus)<>ord(knot)  or   ord(node)<>ord(shin)))=
branch(bus,node,'bij')*(Binv(bus,knot)−Binv(bus,shin)−Binv(node,knot)
+Binv(node,shin))/(1−branch(knot,shin,'bij')*(Binv(knot,knot)
+Binv(shin,shin)−2*Binv(knot,shin)));

Display  Bmatrix,Binv,GSHF,dfactor;
```

Line outage distribution factors ($LODF_{ij,nm}$) are described in Table 9.3.

The branch data for IEEE RTS 24-bus network is provided in Table 9.4. The LODF and GSF coefficients are calculated for IEEE RTS 24-bus (Fig. 9.12) and are given in Tables 9.5 and 9.6, respectively.

9.3 Transmission Network Switching

The idea of optimal transmission switching (OTS) has been broadly investigated in the literature [18]. Opening a set of transmission lines would change the network topology and the line flow patterns. This can be used to relieve the line congestion in the system and reduce the operating costs. Some research works which used the transmission switching as a flexibility tool are listed as follows:

Table 9.3 Line outage distribution factors ($LODF_{ij,nm}$) for three-bus network

Line	m		
ij	n	1	2
2–1	3	1	−1
3–1	2	1	
3–1	3		1
3–2	2	−1	
3–2	3	1	

Table 9.4 Branch data for IEEE RTS 24-bus network

From	To	r(pu)	x(pu)	b(pu)	Rating (MVA)	From	To	r(pu)	x(pu)	b(pu)	Rating (MVA)
1	2	0.0026	0.0139	0.4611	175	11	13	0.0061	0.0476	0.0999	500
1	3	0.0546	0.2112	0.0572	175	11	14	0.0054	0.0418	0.0879	500
1	5	0.0218	0.0845	0.0229	175	12	13	0.0061	0.0476	0.0999	500
2	4	0.0328	0.1267	0.0343	175	12	23	0.0124	0.0966	0.2030	500
2	6	0.0497	0.1920	0.0520	175	13	23	0.0111	0.0865	0.1818	500
3	9	0.0308	0.1190	0.0322	175	14	16	0.0050	0.0389	0.0818	500
3	24	0.0023	0.0839	0.0000	400	15	16	0.0022	0.0173	0.0364	500
4	9	0.0268	0.1037	0.0281	175	15	21	0.0032	0.0245	0.2060	1000
5	10	0.0228	0.0883	0.0239	175	15	24	0.0067	0.0519	0.1091	500
6	10	0.0139	0.0605	2.4590	175	16	17	0.0033	0.0259	0.0545	500
7	8	0.0159	0.0614	0.0166	175	16	19	0.0030	0.0231	0.0485	500
8	9	0.0427	0.1651	0.0447	175	17	18	0.0018	0.0144	0.0303	500
8	10	0.0427	0.1651	0.0447	175	17	22	0.0135	0.1053	0.2212	500
9	11	0.0023	0.0839	0.0000	400	18	21	0.0017	0.0130	0.1090	1000
9	12	0.0023	0.0839	0.0000	400	19	20	0.0026	0.0198	0.1666	1000
10	11	0.0023	0.0839	0.0000	400	20	23	0.0014	0.0108	0.0910	1000
10	12	0.0023	0.0839	0.0000	400	21	22	0.0087	0.0678	0.1424	500

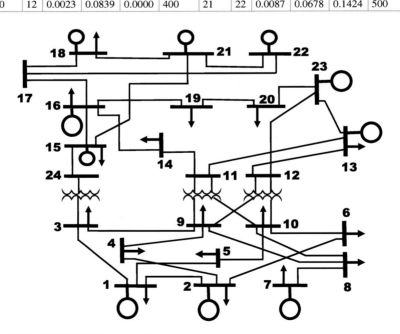

Fig. 9.12 IEEE RTS 24-bus network

Table 9.5 LODF$_{12,\ell}$ calculated for IEEE RTS 24-bus network

$\ell = nm$	1	2	3	4	5	6	8	9	10	11	12	13	14	15	16	17	18	19	20	21
3	0.611																			
4		−0.712																		
5	0.706																			
6		−0.712																		
9			0.291	−0.712			0.054													
10					0.706	−0.712	−0.054													
11								−0.067	0.022											
12								−0.062	0.028											
13										−0.002	−0.010									
14										−0.048	−0.048									
16													−0.048	0.031						
17															−0.009					
18																−0.006				
19															0.043					
20																		0.043		
21														0.009			−0.006			
22																−0.003			0.043	
23											−0.023	−0.016								0.003
24			0.154											−0.154						

Table 9.6 $GSHF_{\ell,m}$ calculated for IEEE RTS 24-bus network

Line	Bus (m)																						
ℓ_{ij}	1	2	3	4	5	6	7	8	9	10	11	12	14	15	16	17	18	19	20	21	22	23	24
2 1	−0.44	0.51	−0.10	0.25	−0.23	0.11	0.01	0.01	0.04	−0.01	0.00	0.00	−0.01	−0.03	−0.02	−0.02	−0.02	−0.02	−0.01	−0.03	−0.02	−0.01	−0.05
3 1	−0.24	−0.22	0.20	−0.09	−0.09	−0.10	−0.03	−0.03	0.01	−0.06	0.00	−0.01	0.02	0.06	0.04	0.05	0.05	0.03	0.02	0.05	0.05	0.02	0.11
4 2	−0.24	−0.27	−0.03	0.37	−0.15	−0.11	0.00	0.00	0.07	−0.06	0.00	0.00	0.00	−0.01	−0.01	−0.01	−0.01	0.00	0.00	−0.01	−0.01	0.00	−0.02
5 1	−0.32	−0.29	−0.10	−0.15	0.38	−0.01	0.01	0.01	−0.05	0.07	0.00	0.00	−0.01	−0.03	−0.02	−0.03	−0.03	−0.02	−0.01	−0.03	−0.03	−0.01	−0.06
6 2	−0.20	−0.22	−0.07	−0.12	−0.07	0.22	0.01	0.01	−0.03	0.05	0.00	0.00	−0.01	−0.02	−0.01	−0.02	−0.02	−0.01	−0.01	−0.02	−0.02	−0.01	−0.04
8 7							−1.00																
9 3	−0.09	−0.07	−0.43	0.03	−0.05	−0.02	0.06	0.06	0.12	−0.01	0.00	0.02	−0.05	−0.12	−0.09	−0.10	−0.11	−0.07	−0.05	−0.11	−0.11	−0.04	−0.24
9 4	−0.24	−0.27	−0.03	−0.63	−0.15	−0.11	0.00	0.00	0.07	−0.06	0.00	0.00	0.00	−0.01	−0.01	−0.01	−0.01	0.00	0.00	−0.01	−0.01	0.00	−0.02
9 8	−0.02	−0.02	0.03	0.03	−0.05	−0.07	−0.50	−0.50	0.07	−0.08	0.00	0.00	0.00	0.01	0.01	0.01	0.01	0.00	0.00	0.01	0.01	0.00	0.02
10 5	−0.32	−0.29	−0.10	−0.15	−0.62	−0.01	0.01	0.01	−0.05	0.07	0.00	0.00	−0.01	−0.03	−0.02	−0.03	−0.03	−0.02	−0.01	−0.03	−0.03	−0.01	−0.06
10 6	−0.20	−0.22	−0.07	−0.12	−0.07	−0.78	0.01	0.01	−0.03	0.05	0.00	0.00	−0.01	−0.02	−0.01	−0.02	−0.02	−0.01	−0.01	−0.02	−0.02	−0.01	−0.04
10 8	0.02	0.02	−0.03	−0.03	0.05	0.07	−0.50	−0.50	−0.07	0.08	0.00	−0.10	0.00	−0.01	−0.01	−0.01	−0.01	0.00	0.00	−0.01	−0.01	0.00	−0.02
11 9	−0.17	−0.17	−0.20	−0.28	−0.12	−0.09	−0.22	−0.22	−0.36	−0.07	0.12	−0.10	0.05	−0.03	−0.01	−0.02	−0.02	−0.02	−0.03	−0.02	−0.02	−0.03	−0.09
11 10	−0.24	−0.24	−0.08	−0.15	−0.31	−0.35	−0.23	−0.23	−0.07	−0.39	0.11	−0.10	0.06	0.00	0.01	0.01	0.01	0.00	−0.01	0.01	0.01	−0.02	−0.03
12 9	−0.18	−0.19	−0.23	−0.29	−0.13	−0.11	−0.23	−0.23	−0.38	−0.08	−0.11	0.12	−0.10	−0.10	−0.08	−0.09	−0.09	−0.05	−0.02	−0.09	−0.09	−0.01	−0.15
12 10	−0.26	−0.25	−0.12	−0.16	−0.33	−0.37	−0.24	−0.24	−0.08	−0.40	−0.11	0.11	−0.08	−0.06	−0.06	−0.06	−0.06	−0.03	−0.01	−0.06	−0.06	0.00	−0.08
13 11	−0.43	−0.43	−0.42	−0.43	−0.43	−0.43	−0.43	−0.43	−0.43	−0.43	−0.63	−0.23	−0.51	−0.41	−0.40	−0.40	−0.40	−0.40	−0.28	−0.40	−0.40	−0.24	−0.41
13 12	−0.40	−0.40	−0.36	−0.40	−0.41	−0.41	−0.41	−0.41	−0.41	−0.41	−0.23	−0.61	−0.26	−0.29	−0.28	−0.28	−0.28	−0.29	−0.28	−0.29	−0.29	−0.29	−0.32
14 11	0.02	0.02	0.14	0.01	0.00	−0.02	−0.02	−0.02	0.00	−0.03	−0.14	0.03	0.63	0.38	0.40	0.40	0.39	0.31	0.24	0.39	0.39	0.19	0.29
16 14	0.02	0.02	0.14	0.01	0.00	−0.02	−0.02	−0.02	0.00	−0.03	−0.14	0.03	−0.37	0.38	0.40	0.40	0.39	0.31	0.24	0.39	0.39	0.19	0.29
16 15	−0.12	−0.12	−0.30	−0.10	−0.09	−0.06	−0.07	−0.07	−0.09	−0.05	0.00	−0.02	0.06	−0.66	0.11	−0.16	−0.29	0.08	0.06	−0.41	−0.31	0.05	−0.52

(continued)

Table 9.6 (continued)

Line ℓ_{ij}	Bus (m)																							
	1	2	3	4	5	6	7	8	9	10	11	12	14	15	16	17	18	19	20	21	22	23	24
21 15	−0.03	−0.03	−0.07	−0.02	−0.02	−0.02	−0.02	−0.02	−0.02	−0.01	0.00	0.00	0.01	−0.15	0.03	0.31	0.45	0.02	0.01	0.57	0.47	0.01	−0.12
21 18	0.02	0.02	0.06	0.02	0.02	0.01	0.01	0.01	0.02	0.01	0.00	0.00	−0.01	0.13	−0.02	−0.27	−0.46	−0.02	−0.01	0.37	0.12	−0.01	0.11
22 17	0.00	0.00	0.01	0.00	0.00	0.00	0.00	0.00	0.00	0.00	0.00	0.00	0.00	0.02	0.00	−0.04	0.01	0.00	0.00	0.06	0.41	0.00	0.02
22 21	0.00	0.00	−0.01	0.00	0.00	0.00	0.00	0.00	0.00	0.00	0.00	0.00	0.00	−0.02	0.00	0.04	−0.01	0.00	0.00	−0.06	0.59	0.00	−0.02
23 12	−0.04	−0.04	0.02	−0.05	−0.05	−0.06	−0.06	−0.06	−0.05	−0.07	0.01	−0.16	0.08	0.13	0.14	0.14	0.14	0.20	0.25	0.14	0.14	0.28	0.09
23 13	0.17	0.17	0.22	0.17	0.16	0.16	0.16	0.16	0.16	0.15	0.13	0.16	0.23	0.31	0.31	0.31	0.31	0.38	0.44	0.31	0.31	0.47	0.27
23 20	−0.13	−0.13	−0.23	−0.12	−0.11	−0.10	−0.10	−0.10	−0.11	−0.09	−0.14	0.00	−0.31	−0.44	−0.46	−0.45	−0.45	−0.58	−0.69	−0.44	−0.45	0.25	−0.36
24 3	−0.15	−0.15	−0.37	−0.13	−0.11	−0.08	−0.08	−0.08	−0.11	−0.06	−0.01	−0.03	0.07	0.18	0.14	0.15	0.16	0.10	0.07	0.17	0.16	0.06	0.35
24 15	0.15	0.15	0.37	0.13	0.11	0.08	0.08	0.08	0.11	0.06	0.01	0.03	−0.07	−0.18	−0.14	−0.15	−0.16	−0.10	−0.07	−0.17	−0.16	−0.06	0.65

- DC-OPF considering $N-1$ contingencies [19]
- Co-optimization of unit commitment and transmission switching with $N-1$ reliability constraints [20]
- Robust transmission switching considering $N-k$ contingencies [21]
- Optimal transmission switching considering short-circuit current limitation constraints [22]
- Probabilistic security analysis of OTS [23]
- Chance-constrained OTS with guaranteed wind power utilization [24]
- Heuristics OTS based on DC-OPF and AC-OPF [25]

In this section, the OTS is solved using GAMS and then we will discuss how this model can be improved and extended. The transmission switching problem is formulated as a MIP model in (9.36).

$$\text{OF} = \sum_{g \in \Omega_G} b_g P_g \tag{9.36a}$$

$$P_{ij} - B_{ij}(\delta_i - \delta_j) \le (1 - \zeta_{ij})M \tag{9.36b}$$

$$P_{ij} - B_{ij}(\delta_i - \delta_j) \ge -(1 - \zeta_{ij})M \tag{9.36c}$$

$$\sum_{g \in \Omega_G^i} P_g + \text{LS}_i - L_i = \sum_{j \in \Omega_\ell^i} P_{ij} : \lambda_i \quad i \in \Omega_B \tag{9.36d}$$

$$-P_{ij}^{\max} \zeta_{ij} \le P_{ij} \le P_{ij}^{\max} \zeta_{ij} \quad ij \in \Omega_\ell \tag{9.36e}$$

$$P_g^{\min} \le P_g \le P_g^{\max} \tag{9.36f}$$

$$\sum_{ij}(1 - \zeta_{ij}) \le N_{\text{sw}} \quad ij \in \Omega_\ell \tag{9.36g}$$

$$B_{ij} = \frac{1}{x_{ij}} \tag{9.36h}$$

$$\zeta_{ij} \in \{0, 1\} \tag{9.36i}$$

where ζ_{ij} is the on/off state of the branch connecting bus i to bus j, and N_{sw} is the number of allowed switching actions in the network.

The operating and congestion cost vs number of switched lines in IEEE 118-bus network (Fig. 9.13) are depicted in Fig. 9.14. The GAMS code for solving the (9.36) is provided in GCode 9.5.

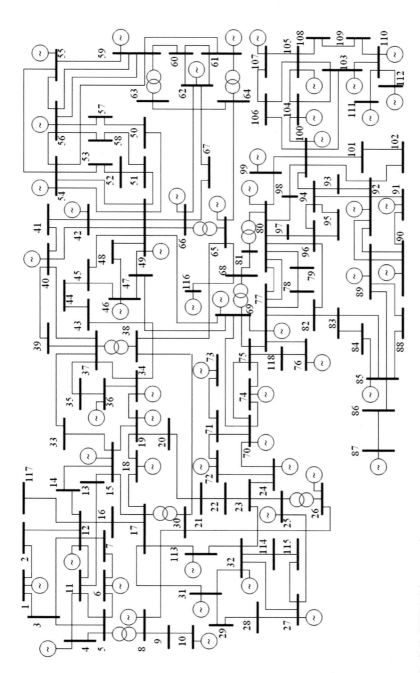

Fig. 9.13   IEEE 118-bus network

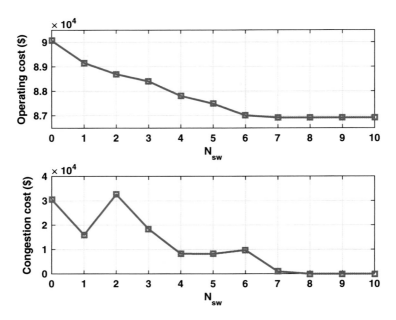

Fig. 9.14 The operating and congestion cost vs number of switched lines in IEEE 118-bus network

GCode 9.5 The OTS GAMS code for IEEE 118-bus network

```
Sets  bus  /1*118/,  slack(bus)  /13/,  conex(bus,node),
GenNo  /Gen1*Gen54/,counter  /c0*c10/;
Scalars       Sbase  /100/;
Alias(bus,totalbus ,node);

Table  GenDatanew(bus,GenNo,*)
                   b        pmin pmax  ;
Table  BusData(bus,*)  buss  characteristics
          Pd  ;
Table  branch(bus,totalbus ,*)
               x          Ilim  ;
branch(bus,totalbus ,'bij')$branch(totalbus ,bus,'x')  =1/branch(bus,
     totalbus ,'x');
conex(bus,node)$branch(bus,node,'x')=yes;
parameter  branch(bus,totalbus ,*),M,NSW, report(counter ,*);
M=smax((bus,node)$conex(bus,node),  branch(bus,node,'bij')*2*pi/3);
Positive  variable  Pg(GenNo);
Variables  Pij(bus,node),delta(bus),ROF;
BINARY  VARIABLE  SW(bus,node);
Equations  const0 , const1 , const2 , const3 , const0A ,
const0B , const0C , const0D , const0E , const0F ;
const0(bus,node)$conex(bus,node)  ..  Pij(bus,node)=e=
branch(bus,node,'bij')*(delta(bus)−delta(node));
const1(bus)..  sum(GenNo$GenDatanew(bus,GenNo,'Pmax'),Pg(GenNo))
```

```
-BusData(bus, 'Pd')/sbase=e=+sum(node$conex(node,bus),Pij(bus,node));
const2       ..   ROF=g= sum((GenNo,bus)$GenDatanew(bus,GenNo,'Pmax'),
GenDatanew(bus,GenNo,'b')*Pg(GenNo)*Sbase);
const0A(bus,node)$conex(bus,node)..
Pij(bus,node)-branch(bus,node,'bij')*(delta(bus)-delta(node))=l= M*(1-SW
     (bus,node));
const0B(bus,node)$conex(bus,node)..
Pij(bus,node)-branch(bus,node,'bij')*(delta(bus)-delta(node))=g=-M*(1-SW
     (bus,node));
const0C(bus,node)$conex(bus,node)..
Pij(bus,node)=l= SW(bus,node)*branch(bus,node,'Ilim');
const0D(bus,node)$conex(bus,node)..
Pij(bus,node)=g=-SW(bus,node)*branch(bus,node,'Ilim');
const0E(bus,node)$conex(bus,node)..SW(node,bus)=e=SW(bus,node);
const0F(bus,node)$conex(bus,node).. Pij(node,bus)=e=-Pij(bus,node);
const3   ..    0.5*sum((bus,node)$conex(bus,node),1-SW(bus,node))=l=NSW;
model BASE /const0,const1,const2/;
model Switching/const1,const2,const0A,const0B,const0C,
                              const0D,const0E,const0F,const3
                                                            /;
Option  Optca=0; Option  Optcr=0;

BusData(bus,'Pd')=1.1*BusData(bus,'Pd');
Pg.lo(GenNo)=sum(bus,GenDatanew(bus,GenNo,'Pmin'))/Sbase;
Pg.up(GenNo)=sum(bus,GenDatanew(bus,GenNo,'Pmax'))/Sbase;
delta.up(bus)=pi/3; delta.lo(bus)=-pi/3; delta.l(bus)=0; delta.fx(slack)
     =0;
Pij.up(bus,node)$conex(bus,node)= 1*branch(bus,node,'Ilim');
Pij.lo(bus,node)$conex(bus,node)=-1*branch(bus,node,'Ilim');
Solve BASE minimizing ROF using lp;
SW.l(bus,node)=1; report('c0','OF')=ROF.l;
report('c0','NSW')=0.5*sum((bus,node)$conex(bus,node),1-SW.l(bus,node));
report('c0','Congestion')=0.5*sum((bus,node)$conex(bus,node),
(-const1.m(bus)+const1.m(node))*Pij.l(bus,node));
loop(counter $(ord(counter)>1),
NSW=ord(counter)-1;
Solve switching  minimizing ROF using mip;
report(counter,'OF')=ROF.l;
report(counter,'NSW')=0.5*sum((bus,node)$conex(bus,node),1-SW.l(bus,node
     ));
report(counter,'Congestion')=0.5*sum((bus,node)$conex(bus,node),
                        (-const1.m(bus)+const1.m(node))*Pij.l(bus,node
                                   ));
);
```

Increasing the demand in a given area or bus (or equivalently losing the generation) might cause congestion and increasing the total operating costs. The OTS can be used to enhance the grid utilization and reduce the line congestions. A simple analysis is conducted as follows:

- The demand at bus i is increased for 20 MW.
- The DC OPF is solved without considering the transmission switching option (LP model).
- The DC OPF along with transmission switching option is solved (MIP model).

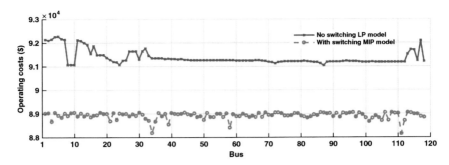

Fig. 9.15 The operating cost vs the connection point of new demand (20 MW) in with/without transmission switching cases

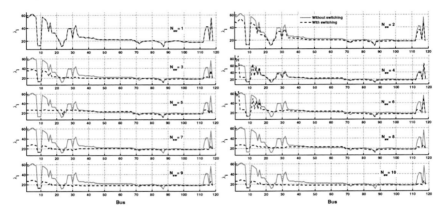

Fig. 9.16 The LMP values (λ_i) in ($/MW h) at different buses in with/without transmission switching cases

The comparison of the total operating costs between the with/without transmission switching cases are shown in Fig. 9.15. The impact of connecting a new demand to different buses would cause different changes in total operating costs. In all cases, using the transmission switching flexibility can reduce the operating costs as shown in Fig. 9.15.

Using the transmission switching might change the LMP values at different buses. The LMP values (λ_i) in $/MW h at different buses in with/without transmission switching cases. As it is shown in Fig. 9.16, when no switching is allowed ($N_{sw} = 0$) then the LMP values of switching and not switching cases are the same. By increasing the number of switchable lines, the LMP values get closer to each other. The generation and branch data of IEEE 118 bus are given in Tables 9.7 and 9.8, respectively.

The developed GAMS code for OTS can be improved to consider the following issues:

Table 9.7 Generation data for 118-bus network

Bus	Unit (g)	b_g ($/MW h)	P_g^{min} (MW)	P_g^{max} (MW)	Bus	Unit (g)	b_g ($/MW h)	P_g^{min} (MW)	P_g^{max} (MW)
1	Gen1	26.2	5.0	30.0	65.0	Gen28	8.3	100.0	420.0
4	Gen2	26.2	5.0	30.0	66.0	Gen29	12.9	80.0	300.0
6	Gen3	26.2	5.0	30.0	69.0	Gen30	15.5	30.0	80.0
8	Gen4	12.9	150.0	300.0	70.0	Gen31	26.2	10.0	30.0
10	Gen5	12.9	100.0	300.0	72.0	Gen32	26.2	5.0	30.0
12	Gen6	26.2	10.0	30.0	73.0	Gen33	37.7	5.0	20.0
15	Gen7	17.8	25.0	100.0	74.0	Gen34	17.8	25.0	100.0
18	Gen8	26.2	5.0	30.0	76.0	Gen35	17.8	25.0	100.0
19	Gen9	26.2	5.0	30.0	77.0	Gen36	12.9	150.0	300.0
24	Gen10	12.9	100.0	300.0	80.0	Gen37	17.8	25.0	100.0
25	Gen11	10.8	100.0	350.0	85.0	Gen38	26.2	10.0	30.0
26	Gen12	26.2	8.0	30.0	87.0	Gen39	10.8	100.0	300.0
27	Gen13	26.2	8.0	30.0	89.0	Gen40	12.9	50.0	200.0
31	Gen14	17.8	25.0	100.0	90.0	Gen41	37.7	8.0	20.0
32	Gen15	26.2	8.0	30.0	91.0	Gen42	22.9	20.0	50.0
34	Gen16	17.8	25.0	100.0	92.0	Gen43	12.9	100.0	300.0
36	Gen17	26.2	8.0	30.0	99.0	Gen44	12.9	100.0	300.0
40	Gen18	26.2	8.0	30.0	100.0	Gen45	12.9	100.0	300.0

(continued)

Table 9.7 (continued)

Bus	Unit (g)	b_g (\$/MW h)	P_g^{min} (MW)	P_g^{max} (MW)	Bus	Unit (g)	b_g (\$/MW h)	P_g^{min} (MW)	P_g^{max} (MW)
42	Gen19	17.8	25.0	100.0	103.0	Gen46	37.7	8.0	20.0
46	Gen20	12.3	50.0	250.0	104.0	Gen47	17.8	25.0	100.0
49	Gen21	12.3	50.0	250.0	105.0	Gen48	17.8	25.0	100.0
54	Gen22	17.8	25.0	100.0	107.0	Gen49	37.7	8.0	20.0
55	Gen23	17.8	25.0	100.0	110.0	Gen50	22.9	25.0	50.0
56	Gen24	13.3	50.0	200.0	111.0	Gen51	17.8	25.0	100.0
59	Gen25	13.3	50.0	200.0	112.0	Gen52	17.8	25.0	100.0
61	Gen26	17.8	25.0	100.0	113.0	Gen53	17.8	25.0	100.0
62	Gen27	8.3	100.0	420.0	116.0	Gen54	22.9	25.0	50.0

Table 9.8 Branch data for 118-bus network

Line	x	Limit	Line	x	Limit	Line	x	Limit	Line	x	Limit	Line	x	Limit	Line	x	Limit
1.2	0.10	1.75	23.25	0.08	5.00	44.45	0.09	1.75	63.64	0.02	5.00	68.81	0.02	5.00	98.100	0.18	1.75
1.3	0.04	1.75	26.25	0.04	5.00	45.46	0.14	1.75	64.61	0.03	5.00	81.80	0.04	5.00	99.100	0.08	1.75
4.5	0.01	5.00	25.27	0.16	5.00	46.47	0.13	1.75	38.65	0.10	5.00	77.82	0.09	2.00	100.101	0.13	1.75
3.5	0.11	1.75	27.28	0.09	1.75	46.48	0.19	1.75	64.65	0.03	5.00	82.83	0.04	2.00	92.102	0.06	1.75
5.6	0.05	1.75	28.29	0.09	1.75	47.49	0.06	1.75	49.66	0.05	10.00	83.84	0.13	1.75	101.102	0.11	1.75
6.7	0.02	1.75	30.17	0.04	5.00	42.49	0.16	3.50	62.66	0.22	1.75	83.85	0.15	1.75	100.103	0.05	5.00
8.9	0.03	5.00	8.30	0.05	1.75	45.49	0.19	1.75	62.67	0.12	1.75	84.85	0.06	1.75	100.104	0.20	1.75
8.5	0.03	5.00	26.30	0.09	5.00	48.49	0.05	1.75	65.66	0.04	5.00	85.86	0.12	5.00	103.104	0.16	1.75
9.10	0.03	5.00	17.31	0.16	1.75	49.50	0.08	1.75	66.67	0.10	1.75	86.87	0.21	5.00	103.105	0.16	1.75
4.11	0.07	1.75	29.31	0.03	1.75	49.51	0.14	1.75	65.68	0.02	5.00	85.88	0.10	1.75	100.106	0.23	1.75
5.11	0.07	1.75	23.32	0.12	1.40	51.52	0.06	1.75	47.69	0.28	1.75	85.89	0.17	1.75	104.105	0.04	1.75
11.12	0.02	1.75	31.32	0.10	1.75	52.53	0.16	1.75	49.69	0.32	1.75	88.89	0.07	5.00	105.106	0.05	1.75
2.12	0.06	1.75	27.32	0.08	1.75	53.54	0.12	1.75	68.69	0.04	5.00	89.90	0.09	10.00	105.107	0.18	1.75
3.12	0.16	1.75	15.33	0.12	1.75	49.54	0.15	3.50	69.70	0.13	5.00	90.91	0.08	1.75	105.108	0.07	1.75
7.12	0.03	1.75	19.34	0.25	1.75	54.55	0.07	1.75	24.70	0.41	1.75	89.92	0.08	10.00	106.107	0.18	1.75
11.13	0.07	1.75	35.36	0.01	1.75	54.56	0.01	1.75	70.71	0.04	1.75	91.92	0.13	1.75	108.109	0.03	1.75
12.14	0.07	1.75	35.37	0.05	1.75	55.56	0.02	1.75	24.72	0.20	1.75	92.93	0.08	1.75	103.110	0.18	1.75
13.15	0.24	1.75	33.37	0.14	1.75	56.57	0.10	1.75	71.72	0.18	1.75	92.94	0.16	1.75	109.110	0.08	1.75
14.15	0.20	1.75	34.36	0.03	1.75	50.57	0.13	1.75	71.73	0.05	1.75	93.94	0.07	1.75	110.111	0.08	1.75
12.16	0.08	1.75	34.37	0.01	5.00	56.58	0.10	1.75	70.74	0.13	1.75	94.95	0.04	1.75	110.112	0.06	1.75
15.17	0.04	5.00	38.37	0.04	5.00	51.58	0.07	1.75	70.75	0.14	1.75	80.96	0.18	1.75	17.113	0.03	1.75
16.17	0.18	1.75	37.39	0.11	1.75	54.59	0.23	1.75	69.75	0.12	5.00	82.96	0.05	1.75	32.113	0.20	5.00

(continued)

Table 9.8 (continued)

Line	x	Limit	Line	x	Limit	Line	x	Limit	Line	x	Limit	Line	x	Limit	Line	x	Limit
17.18	0.05	1.75	37.40	0.17	1.75	56.59	0.13	3.50	74.75	0.04	1.75	94.96	0.09	1.75	32.114	0.06	1.75
18.19	0.05	1.75	30.38	0.05	1.75	55.59	0.22	1.75	76.77	0.15	1.75	80.97	0.09	1.75	27.115	0.07	1.75
19.20	0.12	1.75	39.40	0.06	1.75	59.60	0.15	1.75	69.77	0.10	1.75	80.98	0.11	1.75	114.115	0.01	1.75
15.19	0.04	1.75	40.41	0.05	1.75	59.61	0.15	1.75	75.77	0.20	1.75	80.99	0.21	2.00	68.116	0.00	5.00
20.21	0.08	1.75	40.42	0.18	1.75	60.61	0.01	5.00	77.78	0.01	1.75	92.100	0.30	1.75	12.117	0.14	1.75
21.22	0.10	1.75	41.42	0.14	1.75	60.62	0.06	1.75	78.79	0.02	1.75	94.100	0.06	1.75	75.118	0.05	1.75
22.23	0.16	1.75	43.44	0.25	1.75	61.62	0.04	1.75	77.80	0.05	10.00	95.96	0.05	1.75	76.118	0.05	1.75
23.24	0.05	1.75	34.43	0.17	1.75	63.59	0.04	5.00	79.80	0.07	1.75	96.97	0.09	1.75	12.117	0.14	1.75
															75.118	0.05	1.75
															76.118	0.05	1.75

- The model should be multi-period. The OTS should consider the variation pattern of demand and then determine the optimal switching actions.
- The uncertainty of demand and renewable power generation should be taken into account.
- The current formulation does not ensure the network connectivity. It only tries to satisfy the nodal demand-supply constraint. The resultant system (after switching) might contain some islands.
- The computation burden of the model should be improved to make it applicable for large scale transmission networks.
- The AC power flow constraints should be used for getting closer to reality.
- The unit commitment constraints can be added to the formulation to consider the on/off states of the units as the decision variables.
- Changing the transmission network topology might change the short circuit level on each bus. This should be taken into account for protection issues.
- The current model only considers the intact condition of the network. The contingencies should also be considered.
- The OTS flexibility can be combined with demand response and power flow controller devices.
- The bus splitting can be regarded as a switching action.

References

1. G.A. Orfanos, P.S. Georgilakis, N.D. Hatziargyriou, Transmission expansion planning of systems with increasing wind power integration. IEEE Trans. Power Syst. **28**(2), 1355–1362 (2013)
2. H. Zhang, V. Vittal, G.T. Heydt, J. Quintero, A mixed-integer linear programming approach for multi-stage security-constrained transmission expansion planning. IEEE Trans. Power Syst. **27**(2), 1125–1133 (2012)
3. I. De J. Silva, M.J. Rider, R. Romero, C.A.F. Murari, Transmission network expansion planning considering uncertainty in demand. IEEE Trans. Power Syst. **21**(4), 1565–1573 (2006)
4. J.D. Finney, H.A. Othman, W.L. Rutz, Evaluating transmission congestion constraints in system planning. IEEE Trans. Power Syst. **12**(3), 1143–1150 (1997)
5. R.A. Jabr, Robust transmission network expansion planning with uncertain renewable generation and loads. IEEE Trans. Power Syst. **28**(4), 4558–4567 (2013)
6. S. Haffner, A. Monticelli, A. Garcia, J. Mantovani, R. Romero, Branch and bound algorithm for transmission system expansion planning using a transportation model. IEE Proc. Gener. Transm. Distrib. **147**(3), 149–156 (2000)
7. A.K. Kazerooni, J. Mutale, Transmission network planning under security and environmental constraints. IEEE Trans. Power Syst. **25**(2), 1169–1178 (2010)
8. P. Maghouli, S.H. Hosseini, M. Oloomi Buygi, M. Shahidehpour, A scenario-based multi-objective model for multi-stage transmission expansion planning. IEEE Trans. Power Syst. **26**(1), 470–478 (2011)
9. J.H. Roh, M. Shahidehpour, L. Wu, Market-based generation and transmission planning with uncertainties. IEEE Trans. Power Syst. **24**(3), 1587–1598 (2009)
10. M.J. Rider, A.V. Garcia, R. Romero, Power system transmission network expansion planning using AC model. IET Gener. Transm. Distrib. **1**(5), 731–742 (2007)

11. P. Maghouli, S.H. Hosseini, M.O. Buygi, M. Shahidehpour, A multi-objective framework for transmission expansion planning in deregulated environments. IEEE Trans. Power Syst. **24**(2), 1051–1061 (2009)
12. H. Yu, C.Y. Chung, K.P. Wong, J.H. Zhang, A chance constrained transmission network expansion planning method with consideration of load and wind farm uncertainties. IEEE Trans. Power Syst. **24**(3), 1568–1576 (2009)
13. M.R. Bussieck, A. Pruessner, Mixed-integer nonlinear programming. SIAG/OPT Newsl. Views News **14**(1), 19–22 (2003)
14. R. Romero, A. Monticelli, A. Garcia, S. Haffner, Test systems and mathematical models for transmission network expansion planning. IEE Proc. Gener. Transm. Distrib. **149**(1), 27–36 (2002)
15. L. Baringo, A.J. Conejo, Transmission and wind power investment. IEEE Trans. Power Syst. **27**(2), 885–893 (2012)
16. A.J. Wood, B.F. Wollenberg, *Power Generation Operation and Control*, 2nd edn. (Wiley, New York, 1996)
17. A.J. Wood, B.F. Wollenberg, *Power Generation, Operation, and Control* (Wiley, New York, 2012)
18. E.B. Fisher, R.P. O'Neill, M.C. Ferris, Optimal transmission switching. IEEE Trans. Power Syst. **23**(3), 1346–1355 (2008)
19. K.W. Hedman, R.P. O'Neill, E.B. Fisher, S.S. Oren, Optimal transmission switching with contingency analysis. IEEE Trans. Power Syst. **24**(3), 1577–1586 (2009)
20. K.W. Hedman, M.C. Ferris, R.P. O'Neill, E.B. Fisher, S.S. Oren, Co-optimization of generation unit commitment and transmission switching with n-1 reliability. IEEE Trans. Power Syst. **25**(2), 1052–1063 (2010)
21. T. Ding, C. Zhao, Robust optimal transmission switching with the consideration of corrective actions for n-k contingencies. IET Gener. Transm. Distrib. **10**(13), 3288–3295 (2016)
22. Z. Yang, H. Zhong, Q. Xia, C. Kang, Optimal transmission switching with short-circuit current limitation constraints. IEEE Trans. Power Syst. **31**(2), 1278–1288 (2016)
23. P. Henneaux, D.S. Kirschen, Probabilistic security analysis of optimal transmission switching. IEEE Trans. Power Syst. **31**(1), 508–517 (2016)
24. F. Qiu, J. Wang, Chance-constrained transmission switching with guaranteed wind power utilization. IEEE Trans. Power Syst. **30**(3), 1270–1278 (2015)
25. M. Soroush, J.D. Fuller, Accuracies of optimal transmission switching heuristics based on dcopf and acopf. IEEE Trans. Power Syst. **29**(2), 924–932 (2014)

Chapter 10
Energy System Integration

This chapter provides a solution for energy system integration (ESI) problem in GAMS. The ESI analysis refers to a class of studies which investigate the potential in different energy sectors (water, gas, and electricity) for moving toward a more environmentally friendly and efficient energy supply. The main idea is how to harvest the flexibilities in each energy carrier in a larger framework. In this chapter, the coordination between water desalination systems and power system, gas network-power network, and finally the concept of energy hub is investigated.

10.1 Water-Power Nexus

The water-electricity interdependence is an undeniable issue. Water is used for cooling the power plants. On the other hand, the electricity is used for collecting, treatment, and disposal of water.

The water-energy nexus concept is shown in Fig. 10.1 [1]. The optimization is formulated as follows:

$$\min_{\text{DV}} \text{OF} = \text{TC} + \text{CC} + \text{WC} \tag{10.1a}$$

$$\text{DV} = \left\{ P_{g,t}, P_{c,t}, W_{c,t}, W_{w,t} \right\} \tag{10.1b}$$

$$\text{CT} = \sum_{g,t} a_g P_{g,t}^2 + b_g P_{g,t} + c_g U_{g,t} \tag{10.1c}$$

$$\text{CC} = \sum_{c,t} \alpha_c P_{c,t}^2 + \beta_c P_{c,t} W_{c,t} + \gamma_c W_{c,t}^2 + \zeta_c P_{c,t} + \varsigma_c W_{c,t} + \xi_c U_{c,t} \tag{10.1d}$$

$$\text{CW} = \sum_{w,t} a_w W_{w,t}^2 + b_w W_{w,t} + c_w U_{w,t} \tag{10.1e}$$

© Springer International Publishing AG 2017
A. Soroudi, *Power System Optimization Modeling in GAMS*,
DOI 10.1007/978-3-319-62350-4_10

Water – Energy Nexus schematic

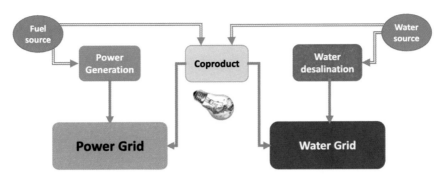

Fig. 10.1 Water-energy nexus concept

Table 10.1 Technical and economical characteristics of thermal units

Unit	a_g ($/MW^2)	b_g ($/MW)	c_g ($)	P_g^{max} (MW)	P_g^{min} (MW)
g_1	0.0002069	−0.1483	57.11	500	0
g_2	0.0003232	−0.1854	57.11	400	0
g_3	0.001065	−0.6026	126.8	400	0
g_4	0.0004222	−0.2119	57.11	350	0

$$P_g^{min} U_{g,t} \leq P_{g,t} \leq P_g^{max} U_{g,t} \tag{10.1f}$$

$$P_c^{min} U_{c,t} \leq P_{c,t} \leq P_c^{max} U_{c,t} \tag{10.1g}$$

$$W_c^{min} U_{c,t} \leq W_{c,t} \leq W_c^{max} U_{c,t} \tag{10.1h}$$

$$W_w^{min} U_{w,t} \leq W_{w,t} \leq W_w^{max} U_{w,t} \tag{10.1i}$$

$$R_c^{min} \leq \frac{P_{c,t}}{W_{c,t}} \leq R_c^{max} \tag{10.1j}$$

$$\sum_g P_{g,t} + \sum_c P_{c,t} = PL_t \tag{10.1k}$$

$$\sum_w W_{w,t} + \sum_c W_{c,t} = WL_t \tag{10.1l}$$

The simulation data are taken from [1] with slight modifications (Tables 10.1, 10.2, and 10.3).

Table 10.2 Technical and economical characteristics of co-production units

Unit	P_c^{max}	P_c^{min}	W_c^{max}	W_c^{min}	R_c^{min}	R_c^{max}	α_c	β_c	γ_c	ζ_c	ς_c	ξ_c
c_1	800	160	200	30	4	9	0.0004433	0.003546	0.007093	−1.106	−4.426	737.4
c_2	600	120	150	23	4	9	0.0007881	0.006305	0.01261	−1.475	−5.901	737.4
c_3	400	80	100	15	4	9	0.001773	0.01419	0.02837	−2.213	−8.851	737.4

Table 10.3 Technical and economical characteristics of water desalination units

Unit	a_w	b_w	c_w	W_w^{max}	W_w^{min}
w_1	0.00182	-0.708	7.374	250	0

GCode 10.1 Water-energy nexus optimization problem

```
Sets    t /t1*t24/,i /p1*p4/, c /c1*c3/, w /w1/;
Table gendata(i,*) generator cost characteristics and limits
        a        b         c       Pmax    Pmin
p1  0.0002069  −0.1483  57.11    500     0
p2  0.0003232  −0.1854  57.11    400     0
p3  0.001065   −0.6026  126.8    400     0
p4  0.0004222  −0.2119  57.11    350     0;
Table Coproduct(c,*)
      Pmax  Pmin  Wmax  Wmin  rmin  rmax  A11        A12       A22       b1      b2      C
c1    800   160   200   30    4     9     0.0004433  0.003546  0.007093  −1.106  −4.426  737.4
c2    600   120   150   23    4     9     0.0007881  0.006305  0.01261   −1.475  −5.901  737.4
c3    400   80    100   15    4     9     0.001773   0.01419   0.02837   −2.213  −8.851
      737.4;
Table waterdata(w,*)
      a         b          c       Wmax    Wmin
w1    1.82E−02  −7.081e−1  7.374   250     0;
Table PWdata(t,*)
      Pd     water
t1    1250   150
t2    1125   130
t3    875    100
t4    750    150
t5    950    200
t6    1440   350
t7    1500   300
t8    1750   200
t9    2000   300
t10   2250   400
t16   2500   550
t17   2125   550
t18   2375   500
t19   2250   400
t20   1975   350
t21   1750   300
t22   1625   250
t23   1500   200
t24   1376   150;
Variables   Of,  p(i,t),TC,CC,Pc(c,t),Wc(c,t),Water(w,t),WaterCost;
Binary variables Up(i,t),Uc(c,t),Uw(w,t);
p.up(i,t)=gendata(i,"Pmax"); p.lo(i,t)=0; Pc.up(c,t)=Coproduct(c,'Pmax');
Pc.lo(c,t)=0; Wc.up(c,t)=Coproduct(c,'Wmax'); Wc.lo(c,t)=0;
Water.up(w,t)=waterdata(w,'Wmax'); Water.lo(w,t)=0;
Equations costThermal,balanceP,balanceW,costCoprodcalc,Objective,
costwatercalc,ratio1,ratio2,EQ1,EQ2,EQ3,EQ4,EQ5,EQ6,EQ7,EQ8;
costThermal..TC=e=sum((t,i), gendata(i,'a')*power(p(i,t),2)+gendata(i,'b')*p(i,t)
```

```
+gendata(i,'c')*Up(i,t));
balanceP(t)  .. sum(i,p(i,t))+sum(c,Pc(c,t))=e=PWdata(t,'Pd');
balanceW(t)  .. sum(w,Water(w,t))+sum(c,Wc(c,t))=e=PWdata(t,'water');
costCoprodcalc.. CC=e=sum((c,t),Coproduct(c,'A11')*power(Pc(c,t),2)
+2*Coproduct(c,'A12')*Pc(c,t)*Wc(c,t) +Coproduct(c,'A22')*power(Wc(c,t),2)
+Coproduct(c,'B1')*Pc(c,t)+Coproduct(c,'B2')*Wc(c,t)+Coproduct(c,'C')*Uc(c,t));
costwatercalc.. WaterCost=e=sum((t,w),  waterdata(w,'a')*power(Water(w,t),2)
+waterdata(w,'b')*Water(w,t) +waterdata(w,'c')*Uw(w,t));
Objective  .. OF=e=TC+CC+WaterCost;
ratio1(c,t)  .. Pc(c,t)=l=Wc(c,t)*Coproduct(c,'Rmax');
ratio2(c,t)  .. Pc(c,t)=g=Wc(c,t)*Coproduct(c,'Rmin');
eq1(w,t)       .. Water(w,t)=l=Uw(w,t)*waterdata(w,'Wmax');
eq2(w,t)..  Water(w,t)=g=Uw(w,t)*waterdata(w,'Wmin');
eq3(c,t)..  wc(c,t)=l= Uc(c,t)*Coproduct(c,'Wmax');
eq4(c,t)..  wc(c,t)=g= Uc(c,t)*Coproduct(c,'Wmin');
eq5(c,t)..  Pc(c,t)=l= Uc(c,t)*Coproduct(c,'Pmax');
eq6(c,t)..  Pc(c,t)=g= Uc(c,t)*Coproduct(c,'Pmin');
eq7(i,t)..  p(i,t) =l=Up(i,t)*gendata(i,"Pmax");
eq8(i,t)..  p(i,t) =g=Up(i,t)*gendata(i,"Pmin");
Model DEDcostbased /all/; Solve DEDcostbased us MInlp min OF;
```

The hourly water output of different plants in water-energy nexus problem is depicted in Fig. 10.2. The hourly power output of different plants in water-energy nexus problem is shown in Fig. 10.3.

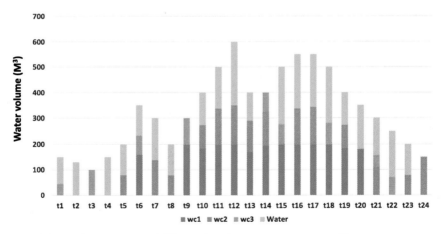

Fig. 10.2 Hourly water output of different plants in water-energy nexus problem

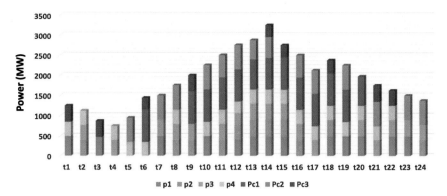

Fig. 10.3 Hourly power output of different plants in water-energy nexus problem

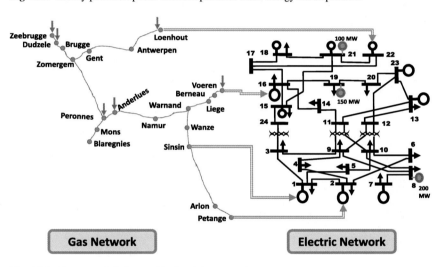

Fig. 10.4 Gas network linkage with electricity network

10.2   Gas-Power Nexus

The interaction of gas network and electricity network is modeled in this section. The electrical network is IEEE RTS 24-bus network which is shown in Fig. 10.4. It is a transmission network with the voltage levels of 138 kV, 230 kV and Sbase = 100 MVA. The branch data for IEEE RTS 24-bus network is given in Table 10.4 [2]. The from bus, to bus, reactance (X), resistance (r), total line charging susceptance (b), and MVA rating (MVA) are specified in this table. The parallel lines in MATPOWER are merged and the resultants are given in Table 10.4. The generation data for IEEE RTS 24-bus network is given in Table 10.5. The data of generating units in this network is inspired by Conejo et al. [3] and Bouffard et al. [4] with some modifications. The slack bus is bus 13 in this network. The wind turbines and the capacities are also shown in Fig. 10.4.

Table 10.4 Branch data for IEEE RTS 24-bus network

From	To	r(pu)	x(pu)	b(pu)	Rating (MVA)	From	To	r(pu)	x(pu)	b(pu)	Rating (MVA)
1	2	0.0026	0.0139	0.4611	175	11	13	0.0061	0.0476	0.0999	500
1	3	0.0546	0.2112	0.0572	175	11	14	0.0054	0.0418	0.0879	500
1	5	0.0218	0.0845	0.0229	175	12	13	0.0061	0.0476	0.0999	500
2	4	0.0328	0.1267	0.0343	175	12	23	0.0124	0.0966	0.2030	500
2	6	0.0497	0.1920	0.0520	175	13	23	0.0111	0.0865	0.1818	500
3	9	0.0308	0.1190	0.0322	175	14	16	0.0050	0.0389	0.0818	500
3	24	0.0023	0.0839	0.0000	400	15	16	0.0022	0.0173	0.0364	500
4	9	0.0268	0.1037	0.0281	175	15	21	0.0032	0.0245	0.2060	1000
5	10	0.0228	0.0883	0.0239	175	15	24	0.0067	0.0519	0.1091	500
6	10	0.0139	0.0605	2.4590	175	16	17	0.0033	0.0259	0.0545	500
7	8	0.0159	0.0614	0.0166	175	16	19	0.0030	0.0231	0.0485	500
8	9	0.0427	0.1651	0.0447	175	17	18	0.0018	0.0144	0.0303	500
8	10	0.0427	0.1651	0.0447	175	17	22	0.0135	0.1053	0.2212	500
9	11	0.0023	0.0839	0.0000	400	18	21	0.0017	0.0130	0.1090	1000
9	12	0.0023	0.0839	0.0000	400	19	20	0.0026	0.0198	0.1666	1000
10	11	0.0023	0.0839	0.0000	400	20	23	0.0014	0.0108	0.0910	1000
10	12	0.0023	0.0839	0.0000	400	21	22	0.0087	0.0678	0.1424	500

The gas network is also shown in Fig. 10.4 which its data is taken from [5]. The technical and economical characteristics of gas nodes are given in Table 10.6. The technical characteristics of gas network are also provided in Table 10.7 [5]. The gas network equations are described in (10.2).

$$GC = \sum_{n,t} c_n Sg_{n,t} \tag{10.2a}$$

$$\sum_m f_{n,m,t} = \sum_m f_{m,n,t} + Sg_{n,t} - \zeta_{g,t} Sd_n - Se_{n,t} \tag{10.2b}$$

$$f_{m,n,t} = C_{m,n} \sqrt{Pr_{m,t}^2 - Pr_{n,t}^2} \quad \text{Passive arcs} \tag{10.2c}$$

$$f_{m,n,t} \geq C_{mn} \sqrt{Pr_{m,t}^2 - Pr_{n,t}^2} \quad \text{Active arcs} \tag{10.2d}$$

$$Sg_n^{min} \leq Sg_{n,t} \leq Sg_n^{max} \tag{10.2e}$$

$$Pr_n^{min} \leq Pr_{n,t} \leq Pr_n^{max} \tag{10.2f}$$

Table 10.5 Generation data for IEEE RTS 24-bus network

Gen	Bus	P_i^{max} (MW)	P_i^{min}	b_i (\$/MW)	Cs_i (\$)	Cd_i (\$)	RU_i (MW h^{-1})	RD_i (MW h^{-1})	SU_i (MW h^{-1})	SD_i (MW h^{-1})	UT_i (h)	DT_i (h)	$u_{i,t=0}$	U_i^0 (h)	S_i^0 (h)
g1	18	400	100	5.47	0	0	47	47	105	108	1	1	1	5	0
g2	21	400	100	5.47	0	0	47	47	106	112	1	1	1	6	0
g3	1	152	30.4	13.32	1430.4	1430.4	14	14	43	45	8	4	1	2	0
g4	2	152	30.4	13.32	1430.4	1430.4	14	14	44	57	8	4	1	2	0
g5	15	155	54.25	16	0	0	21	21	65	77	8	8	0	0	2
g6	16	155	54.25	10.52	312	312	21	21	66	73	8	8	1	10	0
g7	23	310	108.5	10.52	624	624	21	21	112	125	8	8	1	10	0
g8	23	350	140	10.89	2298	2298	28	28	154	162	8	8	1	5	0
g9	7	350	75	20.7	1725	1725	49	49	77	80	8	8	0	0	2
g10	13	591	206.85	20.93	3056.7	3056.7	21	21	213	228	12	10	0	0	8
g11	15	60	12	26.11	437	437	7	7	19	31	4	2	0	0	1
g12	22	300	0	0	0	0	35	35	315	326	0	0	1	2	0

Table 10.6 Technical and economical characteristics of gas nodes

Gas node	Sg_n^{min} (10^6 Scm)	Sg_n^{max} (10^6 Scm)	Sd_n (10^6 Scm)	Pr_n^{min} (bar)	Pr_n^{max} (bar)	c_n \$/MBtu
Anderlues	0	1.20	0.00	0.00	66.20	0.00
Antwerpen	0	0.00	4.03	1.25	80.00	0.00
Arlon	0	0.00	0.22	0.00	66.20	0.00
Berneau	0	0.00	0.00	0.00	66.20	0.00
Blaregnies	0	0.00	15.62	2.08	66.20	0.00
Brugge	0	0.00	3.92	1.25	80.00	0.00
Dudzele	0	8.40	0.00	0.00	77.00	2.28
Gent	0	0.00	5.26	1.25	80.00	0.00
Liege	0	0.00	6.39	1.25	66.20	0.00
Loenhout	0	4.80	0.00	0.00	77.00	2.28
Mons	0	0.00	6.85	0.00	66.20	0.00
Namur	0	0.00	2.12	0.00	66.20	0.00
Petange	0	0.00	1.92	1.04	66.20	0.00
Peronnes	0	0.96	0.00	0.00	66.20	1.68
Sinsin	0	0.00	0.00	0.00	63.00	0.00
Voeren	0	22.01	0.00	2.08	66.20	1.68
Wanze	0	0.00	0.00	0.00	66.20	0.00
Warnand	0	0.00	0.00	0.00	66.20	0.00
Zeebrugge	0	11.59	0.00	0.00	77.00	2.28
Zomergem	0	0.00	0.00	0.00	80.00	0.00

The electrical network equations are described in (10.3).

$$EC = \sum_{g,t} a_g(P_{g,t})^2 + b_g P_{g,t} + c_g + \sum_{i,t} VOLL \times LS_{i,t} + VWC \times P_{i,t}^{wc} \quad (10.3a)$$

$$\sum_{g \in \Omega_G^i} P_{g,t} + LS_{i,t} + P_{i,t}^w - L_{i,t} = \sum_{j \in \Omega_\ell^i} P_{ij,t} : \lambda_{i,t} \quad (10.3b)$$

$$P_{ij,t} = \frac{\delta_{i,t} - \delta_{j,t}}{X_{ij}} \quad (10.3c)$$

$$-P_{ij}^{max} \leq P_{ij,t} \leq P_{ij}^{max} \quad (10.3d)$$

$$P_g^{min} \leq P_{g,t} \leq P_g^{max} \quad (10.3e)$$

$$P_{g,t} - P_{g,t-1} \leq RU_g \quad (10.3f)$$

$$P_{g,t-1} - P_{g,t} \leq RD_g \quad (10.3g)$$

$$0 \leq LS_{i,t} \leq L_{i,t} \quad (10.3h)$$

Table 10.7 Technical characteristics of gas network

Pipe	From	To	Active	$C^2_{m,n}$
L_1	Zeebrugge	Dudzele		9.07027
L_2	Zeebrugge	Dudzele		9.07027
L_3	Dudzele	Brugge		6.04685
L_4	Dudzele	Brugge		6.04685
L_5	Brugge	Zomergem		1.39543
L_6	Loenhout	Antwerpen		0.10025
L_7	Antwerpen	Gent		0.14865
L_8	Gent	Zomergem		0.22689
L_9	Zomergem	Peronnes		0.65965
L_{10}	Voeren	Berneau	1	7.25622
L_{11}	Voeren	Berneau	1	0.10803
L_{12}	Berneau	Liege		1.81405
L_{13}	Berneau	Liege		0.02701
L_{14}	Liege	Warnand		1.45124
L_{15}	Liege	Warnand		0.02161
L_{16}	Warnand	Namur		0.86384
L_{17}	Namur	Anderlues		0.90703
L_{18}	Anderlues	Peronnes		7.25622
L_{19}	Peronnes	Mons		3.62811
L_{20}	Mons	Blaregnies		1.45124
L_{21}	Warnand	Wanze		0.05144
L_{22}	Wanze	Sinsin	1	0.00642
L_{23}	Sinsin	Arlon		0.00170
L_{24}	Arlon	Petange		0.02782

$$P^{\mathrm{wc}}_{i,t} = w_t \Lambda^w_i - P^w_{i,t} \tag{10.3i}$$

$$0 \le P^w_{i,t} \le w_t \Lambda^w_i \tag{10.3j}$$

The overall optimization problem, constraints, and the decision variables are as follows:

$$\min_{DV} OF = EC + GC \tag{10.4}$$

$$DV = \left\{ \begin{matrix} \delta_{i,t}, P_{g,t}, P_{c,t}, W_{c,t}, W_{w,t} \\ \mathrm{Sg}_{n,t}, f_{n,m,t}, \mathrm{Pr}_{n,t} \end{matrix} \right\}$$

(10.2 and (10.3)

The hourly variation pattern of wind generation, electric and gas demand is shown in Fig. 10.5.

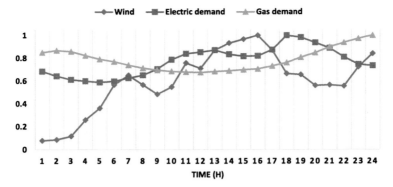

Fig. 10.5 Hourly variation pattern of wind generation, electric and gas demand (pu)

The integrated electricity-gas problem is solved using the GCode 10.2. The total electricity cost EC is 3.9760×10^5. The total gas extraction costs are $GC = \$5.1755 \times 10^5$ and the total costs are $\$9.1515 \times 10^5$. The hourly variation pattern of wind generation, electric and gas demand is shown in Fig. 10.5.

GCode 10.2 Gas-electricity nexus optimization problem

```
Sets    bus   /1*24/ ,slack(bus) /13/,Gen /g1*g12/, t /t1*t24/
          genD(gen) /g1*g2,g5,g7*g11/, genN(gen) /g3,g4,g6,g12/ ;
scalars Sbase /100/   ,VOLL /10000/,VOLW /50/; alias(bus,node);
Sets gn NODES / Anderlues, Antwerpen, Arlon, Berneau, Blaregnies, Brugge, Dudzele,
             Gent, Liege, Loenhout, Mons, Namur, Petange, Peronnes, Sinsin,
             Voeren, Wanze, Warnand, Zeebrugge, Zomergem /
     a PIPES   / L1*L24 /; Alias (gn,gm); set Pnm(a,gn,gm) arc description;
table Ndata(gn,*) Node Data
          slo      sup      Sd      plo  pup  c
* Removed for saving space   ;
set GElink(gn,gen)
/Loenhout      .       g12
Voeren         .       g6
Sinsin         .       g3
Petange        .       g4/;
table AData(a,gn,gm,*) Arc Data
                              act  C2mn;
table GD(Gen,*)  Generating units characteristics;
set GB(bus, Gen) connectivity index of each generating unit to each bus ;
Table BusData(bus,*) Demands of each bus in MW;
Table branch(bus,node,*)   Network technical
              r      x      b      z   limit;
Table DataWDL(t,*)
     w              d              g;
Parameters Wcap(bus),conex(bus,node),SD(gn);
branch(bus,node,'bij')$branch(bus,node,'Limit') =1/branch(bus,node,'x');
conex(bus,node)$(branch(bus,node,'limit')and branch(node,bus,'limit'))=1;
conex(bus,node)$(conex(node,bus))=1; Variables f(a,gn,gm,t),sg(gn,t),pressure(gn,t)

EC, Pij(bus,node,t),Pg(Gen,t),delta(bus,t),lsh(bus,t),Pw(bus,t),pc(bus,t),Gc,OF  ;
Pnm(a,gn,gm)$adata(a,gn,gm,'c2mn')=yes;
Equations const1,const2,const3,const4,const5,const6,CG1,CG2,CG3,CG4,Objective;
const1(bus,node,t)$conex(bus,node).. Pij(bus,node,t)=e=
```

```
branch ( bus , node , ' bij ' ) *( delta ( bus , t )—delta ( node , t ) );
const2 ( bus , t ) .. lsh ( bus , t ) $BusData ( bus , 'pd' )+Pw( bus , t ) $Wcap ( bus )+sum (Gen$GB ( bus , Gen )
,
Pg (Gen , t )—DataWDL ( t , 'd' )*BusData ( bus , 'pd' ) / Sbase=e=
+sum ( node$conex ( node , bus ) , Pij ( bus , node , t ) );
const3 .. EC=e=sum (( bus , GenD , t ) $GB ( bus , GenD ) , Pg (GenD , t )*GD(GenD , 'b' )*Sbase )
+sum (( bus , t ) ,VOLL*lsh ( bus , t )*Sbase$BusData ( bus , 'pd' )+VOLW*Pc ( bus , t )*sbase$Wcap ( bus )
) ;
const4 ( gen , t ) .. pg ( gen , t +1)—pg ( gen , t )=l=GD( gen , 'RU' ) / Sbase ;
const5 ( gen , t ) .. pg ( gen , t −1)—pg ( gen , t )=l=GD( gen , 'RD' ) / Sbase ;
const6 ( bus , t ) $Wcap ( bus ) .. pc ( bus , t )=e=DataWDL ( t , 'w' )*Wcap ( bus ) / Sbase—pw ( bus , t );
Pg . lo (Gen , t )=GD(Gen , 'Pmin' ) / Sbase ; Pg . up (Gen , t )=GD(Gen , 'Pmax' ) / Sbase ;
delta . up ( bus , t )=pi /2; delta . lo ( bus , t )=—pi /2; delta . fx ( slack , t ) =0;
Pij . up ( bus , node , t ) $(( conex ( bus , node )))=1* branch ( bus , node , 'Limit' ) / Sbase ;
Pij . lo ( bus , node , t ) $(( conex ( bus , node )))=—1*branch ( bus , node , 'Limit' ) / Sbase ;
lsh . up ( bus , t )= DataWDL ( t , 'd' )*BusData ( bus , 'pd' ) / Sbase ; lsh . lo ( bus , t )= 0;
Pw. up ( bus , t )=DataWDL ( t , 'w' )*Wcap ( bus ) / Sbase ; Pw. lo ( bus , t )=0;
Pc . up ( bus , t )=DataWDL ( t , 'w' )*Wcap ( bus ) / Sbase ; Pc . lo ( bus , t ) =0; SD( gn )=Ndata ( gn , 'SD' );
CG1( gn , t ) .. sum (Pnm( a , gn ,gm ) , f ( Pnm, t ) )=e=sum (Pnm( a , gm, gn ) , f ( Pnm, t ) )
+sg ( gn , t ) $( Ndata ( gn , 'Sup' )>0)—DataWDL ( t , 'G' )*SD( gn )
—sum (( GenN ) $Gelink ( gn , GenN ) , Pg (GenN , 'b' )*Sbase /35315 );
CG2(Pnm( a , gn ,gm) , t ) $(AData ( a , gn ,gm, 'C2mn' ) AND AData ( a , gn ,gm, 'ACT' )=0)
.. signpower ( f ( Pnm, t ) ,2) =e= AData (Pnm, 'C2mn' )*( pressure ( gn , t )—pressure (gm, t ) );
CG3(Pnm( a , gn ,gm) , t ) $(AData ( a , gn ,gm, 'C2mn' ) AND AData ( a , gn ,gm, 'ACT' )=1)
.. —sqr ( f ( Pnm, t ) ) =g= AData (Pnm, 'C2mn' )*( pressure ( gn , t )—pressure (gm, t ) );
CG4.. Gc =e= sum (( gn , t ) , 35315* ndata ( gn , 'c' )*sg ( gn , t ) $Ndata ( gn , 'Sup' ));
Objective .. OF=e=EC+Gc ; sg . lo ( gn , t ) =0; sg . up ( gn , t )= ndata ( gn , 'sup' );
pressure . lo ( gn , t ) = sqr ( ndata ( gn , 'plo' )); pressure . up ( gn , t ) = sqr ( ndata ( gn , 'pup' ));
f . lo (Pnm( a , gn ,gm) , t ) $(AData ( a , gn ,gm, 'C2mn' )) =
—sqrt (AData ( a , gn ,gm, 'C2mn' )*( pressure . up ( gn , t )—pressure . lo ( gn , t )));
f . up (Pnm( a , gn ,gm) , t ) $(AData ( a , gn ,gm, 'C2mn' ))=
sqrt (AData ( a , gn ,gm, 'C2mn' )*( pressure . up ( gn , t )—pressure . lo ( gn , t )));
f . lo (Pnm( a , gn ,gm) , t ) $(AData ( a , gn ,gm, 'C2mn' ) AND AData ( a , gn ,gm, 'ACT' )=1) =0;
Model overall / all /; Solve overall using nlp min OF;
```

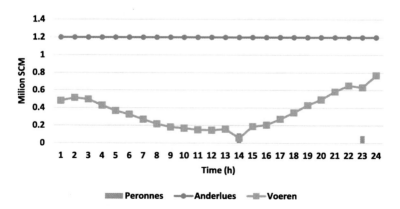

Fig. 10.6 Hourly variation pattern of gas generation from gas sources

Hourly variation pattern of gas generation from gas sources are shown in Fig. 10.6.

The hourly variation pattern of electricity power generation is shown in Fig. 10.7.

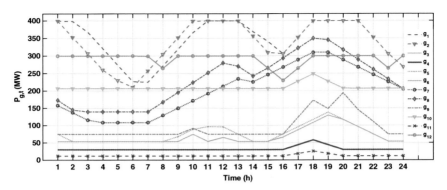

Fig. 10.7 Hourly variation pattern of electricity power generation

The interaction of gas network and electricity sector has been investigated in several works. The impacts of the gas network on security constrained UC is analyzed in [6]. A unified gas and power flow analysis in natural gas and electricity coupled networks can be found in [7]. A robust scheduling model for wind-integrated energy systems with the considerations of both gas pipeline and power transmission contingencies is developed in [8]. The reliability of gas networks and their impacts on the reliability of electricity network is modeled in [9]. The impact of large penetration of wind generation on the UK gas network is analyzed in [10]. One of the recent efficient methods of electricity storage is storing the electricity as gas. This method is also called power to gas or P2G technique [11].

10.3   Energy Hub Concept

The concept of Energy Hub was introduced in [12]. Energy hub may be considered as a virtual box that can convert a set of input energy carriers into a set of energy demands. This box contains several technologies that can store, transfer, or convert different forms of energies to each other. A general example of Energy hub is shown in Fig. 10.8. Different aspects of energy hubs are investigated in the literature such as economic dispatch of energy hubs [13], demand response and energy hub [14], energy hub concept applied on car manufacturing plants [15], and wind power uncertainty modeling in energy hubs [16, 17].

The technologies shown in Fig. 10.8 are explained as follows:

- Combined heat and power (CHP): receives the natural gas (G_t) and converts it into heat (H_t) and electricity (E_t) The CHP economic dispatch problem can be modeled in (10.5) [18]:

$$H_t = \eta_{ge}^{chp} G_t \tag{10.5a}$$

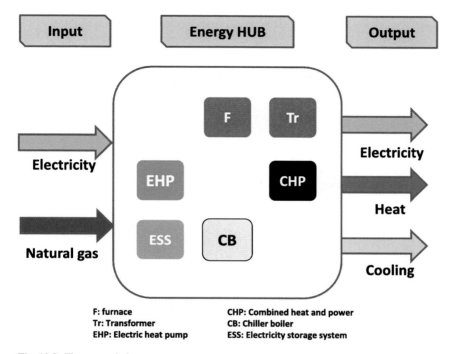

Fig. 10.8 The energy hub concept

$$E_t = \eta_{\mathrm{gh}}^{\mathrm{chp}} G_t \qquad (10.5b)$$

- Electric heat pump (EHP): It is fed by electricity and generates heat demand (H_t) or cool demand (C_t) based on the operating mode. The operation of EHP is mathematically formulated as follows:

$$C_t + H_t = E_t \times \mathrm{COP} \qquad (10.6a)$$

$$H_t^{\min} I_t^h \leq H_t \leq H_t^{\max} I_t^h \qquad (10.6b)$$

$$C_t^{\min} I_t^c \leq C_t \leq C_t^{\max} I_t^c \qquad (10.6c)$$

$$I_t^c + I_t^h \leq 1 \qquad (10.6d)$$

$$I_t^c, I_t^h \in \{0, 1\}$$

The EHP can be in heat or cool generation mode. COP is the coefficient of performance for EHP.

- Chiller boiler (CB): It receives heat and transforms it into cool demand The chiller boiler operation is mathematically formulated as follows:

$$C_t = \eta_{hc} H_t \qquad (10.7)$$

The η_{hc} is the efficiency of heat to cooling conversion for chiller boiler.

- Electricity storage system (ESS): It can store (electricity) and then discharge electricity The ESS operation is mathematically formulated as follows:

$$SOC_t = SOC_{t-1} + (E_t^{ch}\eta_c - E_t^{dch}/\eta_d)\Delta_t \qquad (10.8a)$$

$$E_{min}^{ch} \leq E_t^{ch} \leq E_{max}^{dch} \qquad (10.8b)$$

$$E_{min}^{dch} \leq E_t^{dch} \leq E_{max}^{dch} \qquad (10.8c)$$

$$SOC_{min} \leq SOC_t \leq SOC_{max} \qquad (10.8d)$$

$$I_t^{dch} + I_t^{ch} \leq 1 \qquad (10.8e)$$

$$I_t^{ch}, I_t^{dch} \in \{0, 1\}$$

SOC_t is the state of charge in ESS. $E_t^{ch/dch}$ is for demonstrating the charged and discharged electricity in ESS. The binary variables I_t^{ch}, I_t^{dch} show the charge or discharge mode of ESS at time t.

- Transformer (Tr): It receives electricity and the output is also electricity (with different voltage level)

$$E_t^{out} = \eta_{ee} E_t^{in} \qquad (10.9)$$

- Furnace (F): receives the natural gas and generates the heat demand

$$H_t = \eta_{gh} G_t \qquad (10.10)$$

Three different energy hub configurations will be analyzed to investigate the level of achievable operational flexibility.

10.3.1   Data

The energy hubs which are analyzed in this chapter would have three types of energy demands namely electric, heat, and cooling demand. Different hourly demand and electricity price data for three energy hub configurations are given in Table 10.8.

- The charging and discharging efficiencies ($\eta_{ch/dch}$) of ESS are assumed to be 0.9. The ESS capacity is $SOC_{max} = 600\,MW\,h$ and $SOC_{min} = 120\,MW\,h$. The initial stored energy in ESS is $120\,MW\,h$. The minimum charging and discharging limits are $E_{min}^{ch/dch} = 0$ and the maximum charging and discharging limits are $E_{max}^{ch/dch} = 120\,MW$.
- The transformer efficiency is $\eta_{ee} = 0.98$.
- The CHP efficiencies for gas to electricity is $\eta_{ge} = 0.35$ and for gas to heat is $\eta_{gh} = 0.45$. The CHP capacity is $250\,MW$.

Table 10.8 Different hourly demand and electricity price data for energy hub configurations

Time	D_t^h (MW)	D_t^e (MW)	D_t^c (MW)	λ_t^e \$/MW h
t_1	21.41	52.10	11.51	36.67
t_2	23.21	66.70	13.68	40.41
t_3	26.09	72.20	16.01	38.48
t_4	26.72	78.37	21.42	38.00
t_5	25.59	120.20	21.97	40.24
t_6	26.45	83.48	30.80	38.55
t_7	39.54	110.40	38.94	52.26
t_8	47.28	124.29	46.78	67.34
t_9	52.12	143.61	50.97	70.47
t_{10}	49.13	149.28	48.86	66.20
t_{11}	69.26	154.19	34.77	73.30
t_{12}	61.97	147.30	32.68	60.82
t_{13}	68.04	200.71	27.77	63.15
t_{14}	68.56	174.37	32.02	70.77
t_{15}	56.40	176.54	33.22	63.09
t_{16}	41.32	136.11	34.13	52.53
t_{17}	37.43	108.71	40.78	57.00
t_{18}	25.44	96.90	43.56	49.15
t_{19}	25.66	89.08	51.48	47.47
t_{20}	21.94	82.49	43.15	49.46
t_{21}	22.44	76.93	36.49	53.07
t_{22}	24.63	66.85	27.68	51.60
t_{23}	22.72	47.17	19.14	50.53
t_{24}	22.59	64.67	11.04	36.38

- For EHP, $C_{\text{Max}}^{\text{ehp}} = H_{\text{Max}}^{\text{ehp}} = 500$ MW; The COP is assumed to be 2.5.
- The furnace efficiency η_{gh}^f is 0.9. The furnace capacity is 600 MW.
- The chiller boiler has a capacity equal to 500 MW and the efficiency is $\eta_{\text{hc}} = 0.95$
- The natural gas price is assumed to be constant for different hours and it is equal to $\lambda_t^g = 12\$/\text{MW h}$

10.3.2 Configuration I

This configuration contains transformer, furnace, and chiller boiler as shown in Fig. 10.9.

$$\min \text{OF} = \sum_t \lambda_t^e E_t + \lambda_t^g G_t \tag{10.11a}$$

$$\eta_{\text{ee}} E_t = D_t^e \tag{10.11b}$$

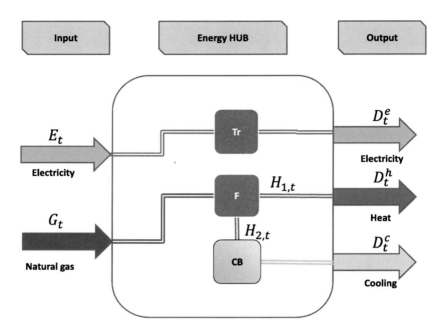

Fig. 10.9 The energy hub configuration I, considering transformer, furnace, and chiller boiler

$$\eta^{f}_{gh} G_t = H_{1,t} + H_{2,t} \tag{10.11c}$$

$$\eta_{hc} H_{2,t} = D^{c}_{t} \tag{10.11d}$$

$$H_{1,t} = D^{h}_{t} \tag{10.11e}$$

The GAMS code for solving the hub configuration I described in (10.11) is given in GCode 10.3.

The developed model in GCode 10.3 is linear and can be solved using any lp solver. The problem is solved and the total operating costs are 1.1327×10^5. The hourly purchased electricity and natural gas in energy hub configuration I are shown in Fig. 10.10.

The output of furnace system will be divided into two streams. The first one will supply the chiller and the second one will directly supply the heat demand. The hourly output of furnace unit in energy hub configuration-I is shown in Fig. 10.11.

GCode 10.3 The optimal operation of energy hub configuration I

```
Set        t        hours             / t1*t24 /
Table  data(t,*)
          Dh     De      Dc     lambda
t1       21.4   52.1    11.5   36.7
t2       23.2   66.7    13.7   40.4
t3       26.1   72.2    16     38.5
t4       26.7   78.4    21.4   38
t5       25.6   120.2   22     40.2
t6       26.4   83.5    30.8   38.6
t7       39.5   110.4   38.9   52.3
t8       47.3   124.3   46.8   67.3
t9       52.1   143.6   51     70.5
t10      49.1   149.3   48.9   66.2
t11      69.3   154.2   34.8   73.3
t12      62     147.3   32.7   60.8
t13      68     200.7   27.8   63.2
t14      68.6   174.4   32     70.8
t15      56.4   176.5   33.2   63.1
t16      41.3   136.1   34.1   52.5
t17      37.4   108.7   40.8   57
t18      25.4   96.9    43.6   49.2
t19      25.7   89.1    51.5   47.5
t20      21.9   82.5    43.1   49.5
t21      22.4   76.9    36.5   53.1
t22      24.6   66.8    27.7   51.6
t23      22.7   47.2    19.1   50.5
t24      22.6   64.7    11     36.4
* ─────────────────────────────────────────
Variable  cost;
Positive  variables     E(t),G(t),H1(t),H2(t);
Scalar
CBmax /500/, eta_ee /0.98/, eta_ghf /0.9/, eta_hc /0.95/;
H2.up(t)=CBmax;
Equations
eq1,eq2,eq3,eq4,eq5;
eq1       ..     cost =e=sum(t,data(t,'lambda')*E(t)+12*G(t));
eq2(t)    ..     eta_ee*E(t)=e=data(t,'E');
eq3(t)    ..     H1(t)=e=data(t,'h');
eq4(t)    ..     eta_ghf*G(t)=e=H1(t)+H2(t);
eq5(t)    ..     eta_hc*H2(t)=e=data(t,'c');
Model Hub / all /;
Solve hub us lp min cost;
```

10.3.3 Configuration II

This configuration contains transformer, furnace, chiller boiler, CHP, and ESS as shown in Fig. 10.12.

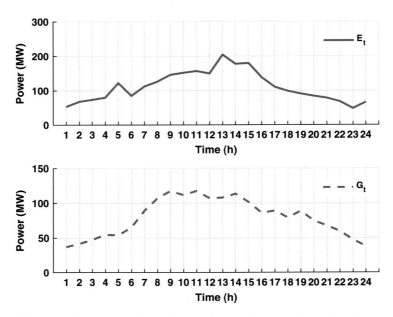

Fig. 10.10 The hourly purchased electricity and natural gas in energy hub configuration I

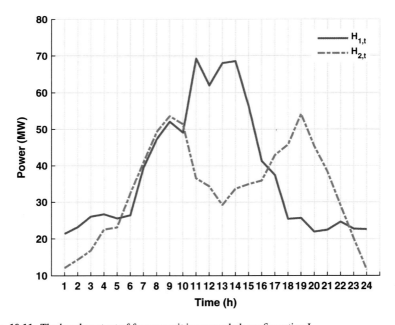

Fig. 10.11 The hourly output of furnace unit in energy hub configuration I

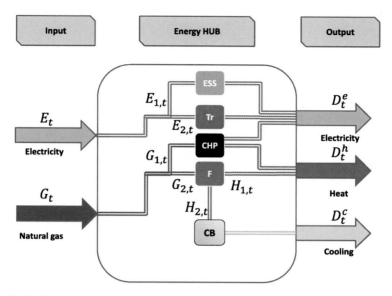

Fig. 10.12 The energy hub configuration II, considering transformer, furnace, chiller boiler, CHP, and electric energy storage

$$\min \text{OF} = \sum_t \lambda_t^e E_t + \lambda_t^g G_t \qquad (10.12\text{a})$$

$$\eta_{ee} E_{2,t} + E_t^{\text{dch}} + \eta_{ge} G_{1,t} = D_t^e \qquad (10.12\text{b})$$

$$E_t = E_{1,t} + E_{2,t} \qquad (10.12\text{c})$$

$$E_{1,t} = E_t^{\text{ch}} \qquad (10.12\text{d})$$

$$\text{SOC}_t = \text{SOC}_{t-1} + (E_t^{\text{ch}} \eta_c - E_t^{\text{dch}}/\eta_d) \Delta_t \qquad (10.12\text{e})$$

$$E_{\min}^{\text{ch}} \leq E_t^{\text{ch}} \leq E_{\max}^{\text{dch}} \qquad (10.12\text{f})$$

$$E_{\min}^{\text{dch}} \leq E_t^{\text{dch}} \leq E_{\max}^{\text{dch}} \qquad (10.12\text{g})$$

$$\text{SOC}_{\min} \leq \text{SOC}_t \leq \text{SOC}_{\max} \qquad (10.12\text{h})$$

$$I_t^{\text{dch}} + I_t^{\text{ch}} \leq 1 \qquad (10.12\text{i})$$

$$I_t^{\text{ch}}, I_t^{\text{dch}} \in \{0, 1\}$$

$$G_t = G_{1,t} + G_{2,t} \qquad (10.12\text{j})$$

$$\eta_{\text{gh}}^f G_{1,t} + H_{1,t} = D_t^h \qquad (10.12\text{k})$$

$$\eta_{\text{gh}}G_{2,t} = H_{1,t} + H_{2,t} \tag{10.12l}$$

$$\eta_{\text{hc}}H_{2,t} = D_t^c \tag{10.12m}$$

The GAMS code for solving the hub configuration II described in (10.12) is given in GCode 10.4.

GCode 10.4 The optimal operation of energy hub configuration II

```
set       t          hours            /t1*t24/

table data(t,*)
     Dh    De    Dc    lambda
t1   21.4  52.1  11.5  36.7
t2   23.2  66.7  13.7  40.4
t3   26.1  72.2  16    38.5
t4   26.7  78.4  21.4  38
t5   25.6  120.2 22    40.2
t6   26.4  83.5  30.8  38.6
t7   39.5  110.4 38.9  52.3
t8   47.3  124.3 46.8  67.3
t9   52.1  143.6 51    70.5
t10  49.1  149.3 48.9  66.2
t11  69.3  154.2 34.8  73.3
t12  62    147.3 32.7  60.8
t13  68    200.7 27.8  63.2
t14  68.6  174.4 32    70.8
t15  56.4  176.5 33.2  63.1
t16  41.3  136.1 34.1  52.5
t17  37.4  108.7 40.8  57
t18  25.4  96.9  43.6  49.2
t19  25.7  89.1  51.5  47.5
t20  21.9  82.5  43.1  49.5
t21  22.4  76.9  36.5  53.1
t22  24.6  66.8  27.7  51.6
t23  22.7  47.2  19.1  50.5
t24  22.6  64.7  11    36.4
Variable cost;
Positive variables
E(t),E1(t),E2(t),G(t),G1(t),G2(t),H1(t),H2(t)
SOC(t),Ec(t),Ed(t)  ;
Binary variables Idch(t),Ich(t);
scalar SOC0 /20/,SOCmax /600/,eta_c /0.9/,eta_d /0.9/,eta_ee /0.98/,eta_ge /0.45/,
eta_gh /0.35/,eta_hc /0.95/,Chpmax /250/,CBmax /500/,Fmax /600/,eta_ghf /0.9/;
SOC0= 0.2*SOCmax;
SOC.up(t)=SOCmax; SOC.lo(t)=0.2*SOCmax; SOC.fx('t24')=SOC0;
Ec.up(t)=0.2*SOCmax; Ec.lo(t)=0;
Ed.up(t)=0.2*SOCmax; Ed.lo(t)=0;
G1.up(t)=Chpmax;
G2.up(t)=Fmax;
H2.up(t)=CBmax;
Equations  eq1,eq2,eq3,eq4,eq5,eq6,eq7,eq8,eq9,eq10,eq11,eq12;
eq1       ..    cost =e=sum(t,data(t,'lambda')*E(t)+12*G(t));
eq2(t)    ..    eta_ee*E2(t)+Ed(t)+eta_ge*G1(t)=e=data(t,'E');
eq3(t)    ..    E(t)=e=E1(t)+E2(t);
eq4(t)    ..    E1(t)=e=Ec(t);
eq5(t)    ..    SOC(t)=e=SOC0$(ord(t)=1)+ SOC(t-1)$(ord(t)>1)+Ec(t)*eta_c-Ed(t)/eta_d
          ;
eq6(t)    ..    Ed(t)=l=0.2*SOCmax*Idch(t);
eq7(t)    ..    Ec(t)=l=0.2*SOCmax*Ich(t);
eq8(t)    ..    Idch(t)+Ich(t)=l=1;
```

```
eq9(t)     ..    G(t)=e=G1(t)+G2(t);
eq10(t)    ..    eta_gh*G1(t)+H1(t)=e=1*data(t,'h');
eq11(t)    ..    eta_ghf*G2(t)=e=H1(t)+H2(t);
eq12(t)    ..    eta_hc*H2(t)=e=data(t,'c');
model Hub2 / all /;
Solve hub2 us mip min cost;
```

The developed model in GCode 10.4 is linear and can be solved using any lp solver. The problem is solved, and the total operating costs are 0.85504×10^5. The hourly purchased electricity and its division between transformer and ESS in energy hub configuration-II are shown in Fig. 10.13. The output of furnace system will be divided into two streams. The first one will supply the chiller and the second one will supply the heat demand. The hourly output of furnace unit in energy hub configuration II is shown in Fig. 10.14. The hourly state of charge, charging, and discharging of ESS in energy hub configuration II is shown in Fig. 10.15.

10.3.4 Configuration III

This configuration contains transformer, furnace, chiller boiler, CHP, and ESS as shown in Fig. 10.16.

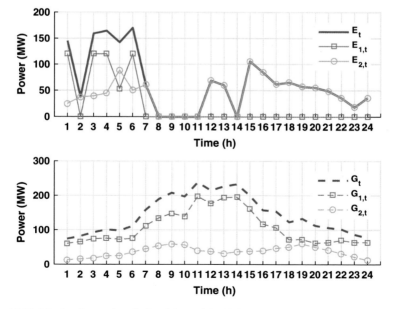

Fig. 10.13 The hourly purchased electricity and its division between transformer and ESS in energy hub configuration II

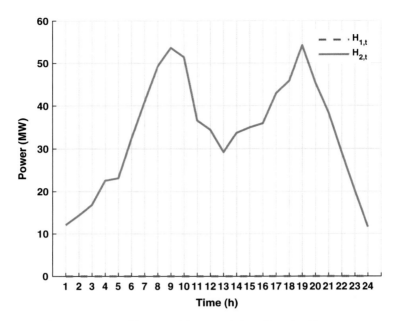

Fig. 10.14 The hourly output of furnace unit in energy hub configuration II

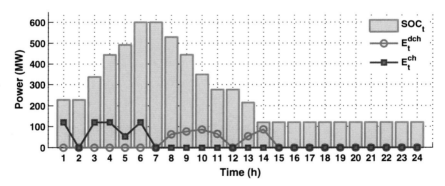

Fig. 10.15 The hourly state of charge (MW h), charging, and discharging of ESS (MW) in energy hub configuration II

$$\min \mathrm{OF} = \sum_t \lambda_t^e E_t + \lambda_t^g G_t \tag{10.13a}$$

$$\eta_{\mathrm{ee}} E_{2,t} + E_t^{\mathrm{dch}} + \eta_{\mathrm{ge}} G_{1,t} = D_t^e \tag{10.13b}$$

$$E_t = E_{1,t} + E_{2,t} + E_{3,t} \tag{10.13c}$$

$$E_{1,t} = E_t^{\mathrm{ch}} \tag{10.13d}$$

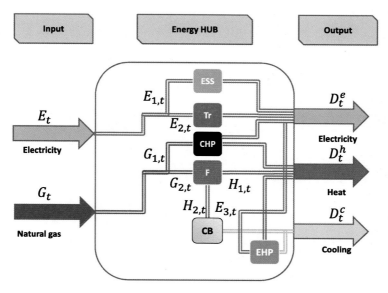

Fig. 10.16 The energy hub configuration III, considering transformer, furnace, chiller boiler, CHP, electric energy storage, and EHP

$$\text{SOC}_t = \text{SOC}_{t-1} + (E_t^{\text{ch}}\eta_c - E_t^{\text{dch}}/\eta_d)\Delta_t \tag{10.13e}$$

$$E_{\min}^{\text{ch}}I_t^{\text{ch}} \leq E_t^{\text{ch}} \leq E_{\max}^{\text{ch}}I_t^{\text{ch}} \tag{10.13f}$$

$$E_{\min}^{\text{dch}}I_t^{\text{dch}} \leq E_t^{\text{dch}} \leq E_{\max}^{\text{dch}}I_t^{\text{dch}} \tag{10.13g}$$

$$\text{SOC}_{\min} \leq \text{SOC}_t \leq \text{SOC}_{\max} \tag{10.13h}$$

$$I_t^{\text{dch}} + I_t^{\text{ch}} \leq 1 \tag{10.13i}$$

$$I_t^{\text{ch}}, I_t^{\text{dch}} \in \{0, 1\}$$

$$G_t = G_{1,t} + G_{2,t} \tag{10.13j}$$

$$\eta_{\text{gh}}G_{1,t} + H_{1,t} + H_t^{\text{EHP}} = D_t^h \tag{10.13k}$$

$$\eta_{\text{gh}}^f G_{2,t} = H_{1,t} + H_{2,t} \tag{10.13l}$$

$$\eta_{\text{hc}}H_{2,t} + C_t^{\text{EHP}} = D_t^c \tag{10.13m}$$

$$C_t^{\text{EHP}} + H_t^{\text{EHP}} = E_{3,t} \times \text{COP} \tag{10.13n}$$

$$H_t^{\min}I_t^h \leq H_t^{\text{EHP}} \leq H_t^{\max}I_t^h \tag{10.13o}$$

$$C_t^{\min}I_t^c \leq C_t^{\text{EHP}} \leq C_t^{\max}I_t^c \tag{10.13p}$$

$$I_t^c + I_t^h \leq 1 \tag{10.13q}$$

$$I_t^c, I_t^h \in \{0, 1\}$$

The developed model in GCode 10.5 is linear and can be solved using any lp solver.

GCode 10.5 The optimal operation of energy hub configuration III

```
Set       t          hours             /t1*t24/

Table  data(t,*)
            Dh    De     Dc    lambda
t1     21.4  52.1   11.5  36.7
t2     23.2  66.7   13.7  40.4
t3     26.1  72.2   16    38.5
t4     26.7  78.4   21.4  38
t5     25.6  120.2  22    40.2
t6     26.4  83.5   30.8  38.6
t7     39.5  110.4  38.9  52.3
t8     47.3  124.3  46.8  67.3
t9     52.1  143.6  51    70.5
t10    49.1  149.3  48.9  66.2
t11    69.3  154.2  34.8  73.3
t12    62    147.3  32.7  60.8
t13    68    200.7  27.8  63.2
t14    68.6  174.4  32    70.8
t15    56.4  176.5  33.2  63.1
t16    41.3  136.1  34.1  52.5
t17    37.4  108.7  40.8  57
t18    25.4  96.9   43.6  49.2
t19    25.7  89.1   51.5  47.5
t20    21.9  82.5   43.1  49.5
t21    22.4  76.9   36.5  53.1
t22    24.6  66.8   27.7  51.6
t23    22.7  47.2   19.1  50.5
t24    22.6  64.7   11    36.4   ;

data(t,'lambda')=0.6*data(t,'lambda');
variable cost;
positive variables E(t),E1(t),E2(t),E3(t),G(t),G1(t),G2(t),H1(t),H2(t)
SOC(t),Ec(t),Ed(t),H_ehp(t),C_ehp(t) ;
Binary variables Idch(t),Ich(t),Ic(t),Ih(t);
scalar SOC0 /20/, SOCmax /600/, eta_c /0.9/, eta_d /0.9/, eta_ee /0.98/
,eta_ge /0.45/ ,eta_gh /0.35/
eta_hc /0.95/, COP /2.5/, H_ehpMax /200/, C_ehpMax /200/,Chpmax /300/,
CBmax /300/,Fmax /300/,eta_ghf /0.9/;
SOC0= 0.2*SOCmax; SOC.up(t)=SOCmax; SOC.lo(t)=0.2*SOCmax; SOC.fx('t24')=SOC0;
Ec.up(t)=0.2*SOCmax; Ec.lo(t)=0; Ed.up(t)=0.2*SOCmax; Ed.lo(t)=0;
C_ehp.up(t)=C_ehpMax; H_ehp.up(t)=H_ehpMax;
G1.up(t)=Chpmax; G2.up(t)=Fmax; H2.up(t)=CBmax; E.up(t)=1000;
Equations
eq1,eq2,eq3,eq4,eq5,eq6,eq7,eq8,eq9,eq10,eq11,eq12,eq13,eq14,eq15,eq16;

eq1       ..     cost =e=sum(t,data(t,'lambda')*E(t)+12*G(t));
eq2(t)    ..     eta_ee*E2(t) +Ed(t)+eta_ge*G1(t)=e=data(t,'E')+E3(t);
eq3(t)    ..     E(t)=e=E1(t)+E2(t);
eq4(t)    ..     E1(t)=e=Ec(t);
eq5(t)    ..     SOC(t)=e=SOC0$(ord(t)=1)+ SOC(t-1)$(ord(t)>1)+Ec(t)*eta_c-Ed(t)/eta_d
          ;
eq6(t)    ..     Ed(t)=l=0.2*SOCmax*Idch(t);
```

```
eq7 ( t )      ..      Ec ( t )=l=0.2*SOCmax*Ich ( t ) ;
eq8 ( t )      ..      Idch ( t )+Ich ( t )=l=1 ;
eq9 ( t )      ..      G ( t )=e=G1 ( t )+G2 ( t ) ;
eq10 ( t )     ..      eta_gh*G1 ( t )+H1 ( t )+H_ehp ( t )=e=l*data ( t , 'h' ) ;
eq11 ( t )     ..      eta_ghf*G2 ( t )=e=H1 ( t )+H2 ( t ) ;
eq12 ( t )     ..      eta_hc*H2 ( t )+C_ehp ( t )=e=data ( t , 'c' ) ;
eq13 ( t )     ..      C_ehp ( t )+H_ehp ( t )=e=E3 ( t )*cop ;
eq14 ( t )     ..      H_ehp ( t )=l=H_ehpMax * Ih ( t ) ;
eq15 ( t )     ..      C_ehp ( t )=l=C_ehpMax * Ic ( t ) ;
eq16 ( t )     ..      Ic ( t )+Ih ( t )=l=1 ;
Model Hub / all / ;
Solve hub us mip min cost ;
```

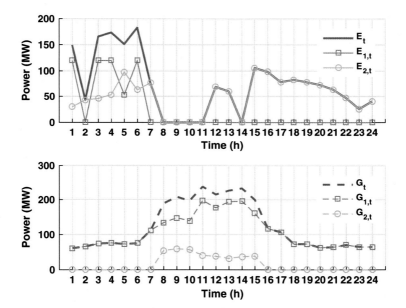

Fig. 10.17 The hourly purchased electricity and its division between transformer and ESS in energy hub configuration III

The problem is solved and the total operating costs are 0.84430×10^5. The hourly purchased electricity and its division between transformer and ESS in energy hub configuration-III are shown in Figs. 10.17, 10.18. The output of furnace system will be divided into two streams. The first one will supply the chiller and the second one will supply the heat demand. The hourly output of furnace unit in energy hub configuration II is shown in Fig. 10.14. The hourly state of charge, charging, and discharging of ESS in energy hub configuration III is shown in Fig. 10.19. The hourly output of EHP in energy hub configuration III is shown in Fig. 10.20.

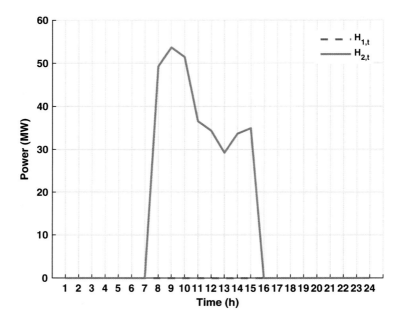

Fig. 10.18 The hourly output of furnace unit in energy hub configuration III

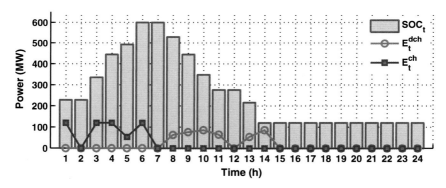

Fig. 10.19 The hourly state of charge (MW h), charging, and discharging of ESS (MW) in energy hub configuration III

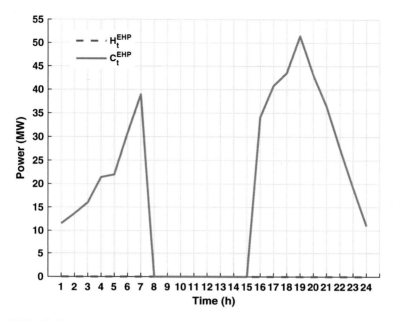

Fig. 10.20 The hourly output of EHP in energy hub configuration III

References

1. A. Santhosh, A.M. Farid, K. Youcef-Toumi, Real-time economic dispatch for the supply side of the energy-water nexus. Appl. Energy **122**, 42–52 (2014)
2. R.D. Zimmerman, C.E. Murillo-Sanchez, R.J. Thomas, Matpower: steady-state operations, planning, and analysis tools for power systems research and education. IEEE Trans. Power Syst. **26**(1), 12–19 (2011)
3. A.J. Conejo, M. Carrión, J.M. Morales, *Decision Making Under Uncertainty in Electricity Markets*, vol. 1 (Springer, Berlin, 2010)
4. F. Bouffard, F.D. Galiana, A.J. Conejo, Market-clearing with stochastic security-part II: case studies. IEEE Trans. Power Syst. **20**(4), 1827–1835 (2005)
5. D. De Wolf, Y. Smeers, The gas transmission problem solved by an extension of the simplex algorithm. Manag. Sci. **46**(11), 1454–1465 (2000)
6. T. Li, M. Eremia, M. Shahidehpour, Interdependency of natural gas network and power system security. IEEE Trans. Power Syst. **23**(4), 1817–1824 (2008)
7. A. Martinez-Mares, C.R. Fuerte-Esquivel, A unified gas and power flow analysis in natural gas and electricity coupled networks. IEEE Trans. Power Syst. **27**(4), 2156–2166 (2012)
8. L. Bai, F. Li, T. Jiang, H. Jia, Robust scheduling for wind integrated energy systems considering gas pipeline and power transmission n-1 contingencies. IEEE Trans. Power Syst. **32**(2), 1582–1584 (2017)
9. J. Munoz, N. Jimenez-Redondo, J. Perez-Ruiz, J. Barquin, Natural gas network modeling for power systems reliability studies. in *2003 IEEE Bologna Power Tech Conference Proceedings*, vol. 4, June 2003, p. 8
10. M. Qadrdan, M. Chaudry, J. Wu, N. Jenkins, J. Ekanayake, Impact of a large penetration of wind generation on the GB gas network. Energy Policy **38**(10), 5684–5695 (2010)

11. H.S. de Boer, L. Grond, H. Moll, R. Benders, The application of power-to-gas, pumped hydro storage and compressed air energy storage in an electricity system at different wind power penetration levels. Energy **72**, 360–370 (2014)
12. M. Geidl, G. Koeppel, P. Favre-Perrod, B. Klöckl, G. Andersson, K. Fröhlich, The energy hub– a powerful concept for future energy systems. in *Third Annual Carnegie Mellon Conference on the Electricity Industry*, Pittsburgh (2007), pp. 13–14
13. S.D. Beigvand, H. Abdi, M. La Scala, A general model for energy hub economic dispatch. Appl. Energy **190**, 1090–1111 (2017)
14. M. Batić, N. Tomašević, G. Beccuti, T. Demiray, S. Vraneš, Combined energy hub optimisation and demand side management for buildings. Energy Build. **127**, 229–241 (2016)
15. K. Kampouropoulos, F. Andrade, Energy hub optimization applied on car manufacturing plants. in *ANDESCON, 2016 IEEE* (IEEE, Piscataway, 2016), pp. 1–4
16. A. Soroudi, B. Mohammadi-Ivatloo, A. Rabiee, Energy hub management with intermittent wind power. in *Large Scale Renewable Power Generation* (Springer, Singapore, 2014), pp. 413–438
17. A. Dolatabadi, B. Mohammadi-Ivatloo, M. Abapour, S. Tohidi, Optimal stochastic design of wind integrated energy hub. IEEE Trans. Ind. Inf. **99**, 1–1 (2017). doi:10.1109/TII.2017.2664101
18. M. Geidl, Integrated modeling and optimization of multi-carrier energy systems. PhD thesis, TU Graz, 2007

Index

© Springer International Publishing AG 2017
A. Soroudi, *Power System Optimization Modeling in GAMS*,
DOI 10.1007/978-3-319-62350-4

Made in the USA
Monee, IL
22 January 2021

f4985a98-abfc-4e02-8380-70b269f5f735R01